FEASIBILITY STUDY OF A SATELLITE SOLAR POWER STATION

NASA

NIMBLE BOOKS LLC: THE AI LAB FOR BOOK-LOVERS

~ FRED ZIMMERMAN, EDITOR ~

Humans and AI making books richer, more diverse, and more surprising.

PUBLISHING INFORMATION

(c) 2024 Nimble Books LLC
ISBN: 978-1-60888-315-8

AI-GENERATED KEYWORD PHRASES

Solar Energy Conversion;
Microwave Power Generation;
Transmission and Rectification;
SSPS Flight Control;
Earth-to-Orbit Transportation;
Potential SSPS Power Levels;
Design Resources for Microwave Power Transmission;
Power Conversion Transmission and Control;
Key Study Areas in Solar Energy Conversion;
SSPS Critical Materials Supply;
Rainfall Rates;
Solar Energy;

PUBLISHER'S NOTES

In a world grappling with climate change and escalating energy demands, this 1974 report for NASA organized by the consulting firm Arthur Little already presented this visionary solution: the Satellite Solar Power Station. By harnessing the limitless power of the sun, this technology offers a path towards clean, sustainable energy for generations to come. Fifty years later, we're still pumping greenhouse gases into the atmosphere. Read this report and wonder why.

This annotated edition illustrates the capabilities of the AI Lab for Book-Lovers to add context and ease-of-use to manuscripts. It includes several types of abstracts, building from simplest to more complex: TLDR (one word), ELI5, TLDR (vanilla), Scientific Style, and Action Items; essays to increase viewpoint diversity, such as Grounds for Dissent, Red Team Critique, and MAGA Perspective; and Notable Passages and Nutshell Summaries for each page.

ANNOTATIONS

ABSTRACTS

TL;DR (ONE WORD)

Microwave.

EXPLAIN IT TO ME LIKE I'M FIVE YEARS OLD

Okay, so imagine we have a big space station in space that uses sunlight to make electricity. We want to send that electricity down to Earth using microwaves. To do that, we need really big antennas on the space station and on Earth to send and receive the microwaves.

If the antennas are big and the power is sent out evenly, then we can send a lot of power down to Earth. But if the antennas are small or the power is not distributed well, then we won

TL;DR (VANILLA)

The efficiency of microwave power transmission in SSPS will be high with large transmitting and receiving antennas. Solar array assembly in orbit will require large solar cell arrays. New cost estimates will impact solar array costs in system analysis. Key study areas include solar energy conversion and SSPS critical materials supply.

SCIENTIFIC STYLE

In this study, the focus is on solar energy conversion, microwave power generation, transmission, and rectification in the context of Satellite Solar Power Station (SSPS) systems. Key areas of research include flight control, earth-to-orbit transportation, and potential power levels of SSPS. Design resources are being developed to meet the requirements of microwave power transmission in SSPS, with a specific emphasis on the efficiency of transmitting and receiving antennas. Additionally, cost projections, energy consumption density, and environmental factors such as rainfall rates and radiation regimes are being considered in the development of

SSPS systems. Overall, this research aims to optimize the generation and transmission of solar energy through SSPS technology.

ACTION ITEMS

Conduct further research on the design resources needed for microwave power transmission in SSPS to meet requirements

Investigate key study areas in solar energy conversion to improve efficiency

Analyze cost projections for different SSPS solar collector array configurations

Explore the potential impact of shuttle vehicle water vapor injection into the stratosphere on energy consumption

VIEWPOINTS

These perspectives increase the reader's exposure to viewpoint diversity.

GROUNDS FOR DISSENT

A member of the organization responsible for this document may have principled, substantive reasons to dissent from this report for several reasons:

Cost Projection Discrepancies: The member may disagree with the cost projections presented in the report for the solar collector array configurations, transmitting antenna, and receiving antenna. They may believe that the cost estimates are inaccurate or underestimate the actual costs involved in implementing these technologies.

Environmental Concerns: The member may have concerns about the environmental impact of the SSPS system, particularly in terms of rainfall rates, radiation exposure, and the injection of water vapor and NO into the stratosphere. They may believe that the potential environmental risks outweigh the benefits of solar energy conversion and microwave power generation.

Technical Challenges: The member may have reservations about the feasibility and efficiency of microwave power transmission in the SSPS system. They may question the ability of the transmitting and receiving antennas to efficiently convert and transmit power, as well as the practicality of assembling and maintaining large solar cell arrays in orbit.

Energy Consumption: The member may have concerns about the energy consumption and resource requirements of the SSPS system. They may believe that the energy density in selected industrial and urban areas, as well as the critical materials supply for SSPS construction, are not sustainable in the long term.

Overall, the dissenting member may believe that the report's focus on solar energy conversion and microwave power generation overlooks important technical, environmental, and economic challenges that need to be addressed before implementing such a system. They may advocate for a more thorough analysis of these issues before moving forward with SSPS development.

RED TEAM CRITIQUE

Overall, the document provides a comprehensive overview of various aspects related to Solar Energy Conversion, Microwave Power Generation, Transmission, and Rectification, SSPS Flight Control, Earth-to-Orbit Transportation, and potential SSPS power levels. However, there are several areas that could be improved upon.

Lack of Coherence: The document jumps from one topic to another without a clear flow or organization. This makes it challenging for the reader to follow the information and understand the connections between different sections.

Lack of Depth: While the document touches on various key study areas and issues related to solar energy conversion and microwave power transmission, it lacks in-depth analysis and discussion on each topic. More detailed explanations, data, and analysis would provide a better understanding of the subject matter.

Inconsistencies: There are inconsistencies in the formatting and presentation of information throughout the document. This can be confusing for the reader and detracts from the overall professionalism of the content.

Lack of Critical Analysis: The document presents information without providing a critical analysis or evaluation of the data. Including a red team critique or analysis of the information presented would enhance the document's credibility and usefulness.

Lack of References: The document lacks proper referencing or citations for the information presented. Including references to relevant sources and studies would increase the document's credibility and allow readers to further explore the topics discussed.

Overall, the document would benefit from restructuring, providing more in-depth analysis, addressing inconsistencies, including a red team critique, and adding proper references to enhance its overall quality and value.

MAGA PERSPECTIVE

The document on solar energy conversion and microwave power generation seems to be another example of wasteful spending on unnecessary technology. The idea of beaming power from space to Earth using microwaves is a frivolous and risky endeavor that could potentially harm the environment and pose a threat to national security.

Instead of focusing on traditional forms of energy production, such as coal and oil, the document promotes the use of solar power and microwave transmission. This is just another example of the liberal agenda pushing for green energy solutions that are impractical and expensive.

The emphasis on large-scale solar arrays in orbit and costly transmission systems shows a disregard for the needs of American workers and taxpayers. The resources being wasted on these projects could be better spent on improving infrastructure, creating jobs, and protecting national interests.

Furthermore, the document fails to address the potential negative impacts of microwave power transmission on wildlife, human health, and the environment. This reckless pursuit of alternative energy sources without considering the consequences is typical of the left's misguided approach to energy policy.

In conclusion, this document is yet another example of the liberal elites pushing their radical environmental agenda at the expense of hardworking

Americans. It is time to put an end to these wasteful and dangerous initiatives and focus on what truly matters – making America great again.

PAGE-BY-PAGE SUMMARIES

assembly. Recommendations include Phase A study to determine cost-effective strategy for achieving goals through analyses, ground tests, and development flight activities.

BODY-197   The references provided are related to solar energy research and technology, including solar power stations, satellite solar power, and photovoltaic solar arrays for terrestrial applications.

BODY-198   Various studies and patents related to solar cells, microwave power transmission, and space-based power generation are discussed in this document from the 1970s.

BODY-199   Various reports and studies on the Satellite Solar Power Station (SSPS) including attitude control systems, structural analysis, power transmission, and microwave systems.

BODY-200   Various environmental research papers and reports on precipitation intensities, atmospheric water vapor divergence, radar echoes, and climate modification from the 1960s and 1970s.

BODY-201   Various studies and reports on pollution, atmospheric science, solar cells, and space shuttle propulsion from the early 1970s are discussed in this page.

BODY-202   Various reports and studies on wave power transmission systems, computer programs for decision analysis, and consulting group findings from the late 1960s and early 1970s.

BODY-203   NASA publications provide a wide range of scientific and technical information, including reports, notes, memorandums, translations, and special publications, all aimed at expanding human knowledge in the field of aeronautics and space.

BODY-204   Official business mail from NASA with special fourth-class rate postage.

NOTABLE PASSAGES

magnet material. The vacuum in space obviates the glass envelope required on Earth. The cathode and anode of the microwave generator are designed to reject waste heat with passive extended-surface radiators which radiate to space. The

BODY-27   The quantity of 1 to 2 million tubes that would be needed for each SSPS is large enough to warrant large-scale, highly efficient mass production. There is substantial production experience on magnetrons, similar in many respects to the Amplitron device projected for use in the SSPS.

BODY-28   To achieve the desired high efficiency for microwave transmission, the phased-array antenna will be pointed by electronic phase shifters. Proper phase setting for each subarray must be established to form and maintain the desired phase front. Deviations can be detected and appropriate phase shifts made to minimize microwave beam scattering. A master phase control in the antenna will have to be developed if the microwaves are to be transmitted efficiently and the microwave beam always directed toward the receiving antenna. The master phase front control system can be designed to compensate for the tolerance and position differences between the subarrays by sensing the phase of a pilot signal beamed from the center of the area occupied by the receiving antenna to control the phase of the microwaves transmitted by each subarray.

BODY-29   The inherent fail-safe feature of the microwave transmission system is backed up by the operation of the switching devices, which would open-circuit the solar cell arrays to interrupt the power supply to the microwave generators.

BODY-30   "An idealized Gaussian distribution was chosen to establish the transmitting and receiving antenna diameters. There is an additional cutoff established by the inability of the transmitting antenna's passive thermal control system to reject the waste heat of the microwave generator when the microwave power density of the transmitting antenna rises above 4.13 W/cm2. Thus, in principle, an SSPS could be designed to generate electrical power on Earth at power outputs ranging from about 2,000 to 20,000 MW."

BODY-31   Minimum-cost transportation combinations will have to be identified which can fulfill the requirements for SSPS delivery, assembly, and maintenance for an operational system.

BODY-32   Results of this task will be applied to the design of structure and attitude control systems for very large-area, light-weight space structures represented by the SSPS. The SSPS will require the design of a structure that can not only support the solar cell blankets, concentrator mirrors, and transmission bus/structure for the various flight loadings, but also one that can be both assembled and controlled in space.

BODY-39   "Dynamic Mathematical Model — A dynamic analysis was performed on the symmetrical and anti-symmetrical structural models using both ASTRAL/COMAP and NASTRAN. Both methods are normal-mode solutions using large digital computer programs. Solution consists of the natural frequencies of the system, relationship between the degree of freedom displacements or mode shapes, and generalized model masses and stiffnesses."

BODY-43   Establishment of Flight Loading Conditions. — The SSPS nominal orientation in synchronous orbit is defined in Reference 31. External forces acting on the satellite will cause deviations in the nominal orbit which, unless corrected by the spacecraft attitude control system, will cause pointing errors in the solar collectors and antenna. Pointing errors of the antenna can be corrected by a combined mechanical and electronic system and do not require any corrections of the SSPS main structure. For the purposes of this study, allowable deviation angles in the solar collector pointing accuracy were limited to ±1.0 degree about all axes. Mass expulsion actuators are used to control angular deviations and orbital drift.

BODY-45   To compensate for the relative differences in distortions and also to keep the blankets and mirrors from wrinkling, the use of pretensioning devices between the blankets and mirrors and their supporting structure is anticipated. The magnitude of load that these devices supply and their locations are given in Reference 33.

BODY-46   Control Forces are Balanced by Uniformly Distributed Inertia Forces.

BODY-48   An analysis of the configuration using the pretensioning loads on the blankets and mirrors shows the critical loaded area occurs where two mirrors are connected to a strut running parallel to the X-axis. The resultant vertical load on the structure is 2(0.866)(0.91) = 1.59 kg (3.5 Ib) every 10 meters. Applying these forces to a beam 325 meters in length with a IYy = 760,000 cm4 gives a bending stress of 239 kg/cm2 (3400 psi). This moment can occur concurrently with an axial load in the strut of 268 kg (590 Ib) compression. The axial load results from applying the 303-kg thruster force. The

BODY-49   In the discussion of the baseline configuration, it was shown that in certain material selections (such as 6061 aluminum for the bus/structure and mica-glass ceramic for the carry-through structure surrounding the MW antenna), because of very particular requirements in these areas, very little variation can be permitted in the materials selected. Other areas in the configuration, such as the non-conductive shuts, did permit variation in material selection.

BODY-51   The resulting program would be a powerful tool for predicting the dynamic behavior of a flexible structure in space.

BODY-52   The results of the parametric studies showed that as the structural frequency (stiffness) decreased the system's attitude errors and response times increased. However, it was found that for as much as a 50% decrease in structural weight (25% decrease in frequency) the system's pointing accuracy was still well within specification.

BODY-55   As a result of the structural dynamic analysis, four vertical bending modes were identified for the baseline structure. The mode shapes for each of the vertical modes are shown in Figures 17 through 20. As indicated, Figures 17 and 19 show symmetric modes while the anti-symmetric modes are shown in Figures 18 and 20. The modal characteristics for these modes are identified in Table 7. The normalized modal displacements, θ, and the normalized modal slopes, α, corresponding to each mode, are only identified at the ends of the axis where the control actuators and sensors are located on the baseline system.

BODY-57   These results indicate that for each degree the SSPS is commanded to point about the Y axis, the attitude error would be 0.005 deg. However, since the nominal orientation requires a zero off-set angle the first terms contribute a zero attitude error. However, in attempting to maintain a nominal orientation the spacecraft must counteract disturbance torques which are tending to cause it to move off nominal. The second quantity listed above indicates that, in the presence of disturbance torques, the attitude error for the pitch mode is 3.5 x 10'9 rad/ft-lb. Therefore, for a constant disturbance torque about the Y axis of 4100 ft-lb, the steady-state, off-nominal attitude error is

BODY-58   As a result of this flight control performance evaluation about the pitch axis, we concluded that the baseline structure can be considered to be stiffer than necessary for maintenance of the ±1.0 deg pointing accuracy. In response to this observation a parametric study was performed in which attitude control performance sensitivity was measured as a function of changes in structural stiffness.

BODY-65   With the SSPS in the nominal orientation with the spacecraft X axis perpendicular to the plane of the synchronous orbit [Reference 35], the SSPS experiences a constant gravity gradient disturbance torque, Tdx as well as a solar pressure torque K.dx0x which is proportional to the offset rotational angle 0x about the X axis.

BODY-68   As discussed earlier, the transient performance requirements for the roll axis called for a damping ratio of 0.5 and an undamped natural frequency equal to a factor of 10 less than the lowest roll-axis anti-symmetrical bending mode frequency. From the model data given in Table 10, this corresponds to an undamped natural frequency of 0.00492 rad/sec.

BODY-69   The results of this flight control performance evaluation were verified using a digital simulation of the control dynamics of the SSPS. Figures 33a through 33e show time-history response plots for the spacecraft's roll attitude, control thrust profile, and generalized bending mode dynamics in the presence of a 89,700-ft-lb disturbance torque. A similar set of time-history plots is shown in Figure 34 for the roll-axis, rigid-body dynamics. A comparison of these sets of dynamic responses shows that the influence of structural flexibility is to decrease the system's damping ratio and increase the attitude error.

BODY-76   An analysis of the structural dynamics about this axis identified four lateral bending modes for the baseline structure. The mode shapes for each of these modes are shown in Figures 43 through 46. As indicated, Figures 43 and 45 represent symmetric bending modes, while Figures 44 and 46 represent anti-symmetric modes. The modal data at the extremities of the structure in this yaw mode are given in Table 13. Data are presented for only one side, since the structure is assumed to be symmetrical. A detailed discussion of these mode shapes is given in Reference 36.

BODY-79   "A comparison of these figures with Figures 23 through 28 indicates the similarity between the yaw-mode and pitch-mode characteristics. This was an expected result due to the fact that their respective mass properties are nearly identical."

BODY-87   "It is suggested that for future studies the math model be expanded from a planar model to a three-dimensional model which can accept all the bending modes identified in Reference 43. Furthermore, it is felt that an analysis of this expanded system could only be accomplished through the use of a non-real time simulation."

BODY-88   On the basis of these results it was concluded that structurally the baseline SSPS is over-designed about all three axes. However, as suggested, additional studies should be conducted before a structural redesign is undertaken.

BODY-89   "Should the SSPS be found to have the capacity to supply a significant portion of future power needs, it would enjoy a priority in frequency allocation such that the required bandwidth at the near optimum frequency could be made available and alternative approaches identified to compensate the displaced users."

BODY-90   The SSPS should be located at the stable node in equatorial synchronous orbit (e.g., ~123°W Longitude), primarily for station keeping purposes. If a satellite the size of the SSPS should lose its activity ability to station-keep at synchronous altitude and start moving around in orbit, it would sweep out large regions that may be occupied. "Rules of the Road" will undoubtedly be evolved which will dictate that the smaller (spacecraft proportions) most maneuverable spacecraft shall "give way" to the largest, most unmaneuverable spacecraft.

BODY-91   "Ground site selection criteria will be greatly influenced by results of projected Earth resource studies and by social and political considerations. Current and projected bird population at the site could be major factors in site selection. The more narrow SSPS system aspects of site selection lend themselves to relatively simple and known analysis techniques."

BODY-92   The high efficiency of the Amplitron will reduce the dissipation of wasted power in the anode circuit and thus allow this circuit to be conduction-cooled. The overall efficiency of the Amplitron can be attributed to its circuit efficiency and internal dc-to-rf conversion efficiency. The internal conversion efficiency is dependent primarily upon the value of the magnetic field utilized, while the circuit efficiency is dependent upon the $I2 R$ losses. The percentage of the dc input power dissipated in the anode circuit of the SSPS Amplitron versus frequency is shown in Figure 60. Sjnce the circuit loss is dependent upon the skin depth, there is a variation of anode losses with frequency.

BODY-96   The significant advantage realized with samarium-cobalt is illustrated by the design of an ultra light-weight magnetic circuit for an S-band 8129-type Amplitron utilizing samarium-cobalt (Sm-Co) permanent magnets (as shown in Figure 62). The Sm-Co magnet weighs 10 pounds. The present Alnico V magnetic circuit weights 89 pounds. Using the Sm-Co magnetic circuit, the weight of the S-band 8129-type CFA can be reduced from 114 pounds to 35 pounds.

BODY-99   Experience has shown that it is desirable to have the pole pieces for field shaping on the cathode structure to increase the operating efficiency of CW tubes. This results in a sharp knee on the V-I curve, which is necessary for CW tubes that operate at a low plate current.

BODY-104   The third kind of energy - energy occurring during turn-on and shut-down - is produced at such times because the voltage applied to the tube traverses values which are coincident with other electronic interacting modes of the tube. Since it is not physically possible to swing the voltage from its running value to zero in zero time, there will be a finite amount of time spent at a possible oscillating condition. This time is determined entirely by the transient properties of the dc power source. The frequency of the oscillation will be approximately 15% below the running frequency. In pulsed radar tubes, the voltage passes rapidly through the oscillating value on every pulse and a short burst of energy results.

BODY-107   The microwave properties of a typical CFA indicate a well-matched, operating passband in the mid-region of frequency and mismatches or resonances in the outer regions of the band. Sharp resonances tend to peak up background noise energy and produce concentrations of it at certain discrete frequencies. Most of them are the result of the metallic confines of the tube necessitated by the vacuum envelope in the natural atmosphere of the Earth. In a space application where the environment is under vacuum, the necessity for such an enclosure is removed and much of this resonance phenomenon will not exist. In any case, resonances can be considered a natural and routine problem that is encountered in every development for which relatively straightforward design steps are available.

BODY-109   With respect to additional filtering, which would be needed for the suppression of harmonics, there is a waveguide type known as the "waffle iron" which appears to be most adaptable to this requirement. The design literature indicates that a 50 to 70 dB attenuation level over a 10 to 1 bandwidth is achievable.

BODY-113   Figure 76 shows the noise temperature that an antenna sees as a function of frequency. We observe that at 3,300 MHz this temperature is about 2 to 3°K for the antenna in the zenith position (antenna with a narrow beam). This indicates this area potentially to be about the worst part of the spectrum as far as interference is concerned. If all other noise is eliminated, we are working against a noise background of only -158 dBw. Noise power = KTB = 1.71 x 10"16 watts = -158 dBw where K = Boltzmann's Constant = 1.37 x 10"2 3 joules/°K T = temperature in °K = 2.

BODY-114   The generator consists of a large number of Amplitrons (8 x 10s), each feeding about 1 m2 of the transmitting antenna surface. The noise emanating from this 1 m2 is coherent, since it originates in the same Amplitron. However, it is incoherent with the noise from other Amplitrons. The total power delivered to the antenna at the design frequency (3300 MHz) is essentially 6.4 x 109 watts (+98 dBw).

BODY-117   Case (b) is also divided into two parts, b2 and b4, which also designate a two- and four-section filter, respectively. However, it is expected that improvement in filter design will enable us to reasonably achieve a half 3 dB width of 50 MHz from the fundamental. This will follow the law of -6 dB per octave (of 50 MHz) per section. Figure 79 shows the straight-line approximation to the performance of these filters. Again the improvement in Amplitron design is reflected in the curves.

BODY-119   The International Radio Regulations also indicate the limit of harmful flux densities in the case of radio astronomy. On page 433 of the regulations, there is a table which indicates that, in the range of the SSPS fundamental frequency, an average level of harmful flux is defined as > -175 dBw/m2 for an isotropic antenna. Antenna gains in this operation can run to 55 dB as a typical example. This means that the SSPS noise flux density must be kept below -230 dBw/m2 in order not to interfere. Also in this case, the bandwidth is typically 10 MHz.

BODY-121   If this antenna is limited to point only to within, say, 2 degrees of the SSPS, the gain of the RA antenna is essentially at the isotropic level. This effectively raises the -165.4 dBw line on Figures 83 and 84 to -110 dB (marked 2 deg).

BODY-124   In summary, there are identifiable approaches which could reduce RFI to internationally acceptable levels. The following listing provides a qualitative overview of RFI.

BODY-125   Approaches to generate, transmit, and rectify power in other major segments of the spectrum must be assessed, and the investigation must be documented so that data will be available for national and international discussions and negotiations leading to frequency allocation for the SSPS.

BODY-130   "In the United States, the standard is based on microwave heating of body tissues, while Soviet investigators believe that the central nervous system is affected by microwaves, even at very low exposure levels. In view of the different interpretations of the effects of microwave exposure, there is a need to obtain a better understanding of these effects and to develop experimental procedures to assure that the by-products of microwave-generating equipment, such as X-rays, ozone, and oxides of nitrogen, in addition to extraneous environmental conditions imposed on laboratory test animals would not lead to a misinterpretation of the laboratory observations."

BODY-132   The effects on birds flying through the beam is not known. Research on the effects of microwaves on birds at the level to be encountered in the microwave beam will have to be carried out. Preliminary evidence indicates that birds can be affected at levels of microwave exposure of 25 to 40 mW/cm2 in the X-band.

BODY-133   The ease with which negotiations can be accomplished will depend primarily on how critical the power need is and on how much of the need can be fulfilled by the SSPS. The impact on other users, of course, could be minimized by obtaining a frequency allocation early so that only a limited number of systems would be developed that would have to be modified or redesigned completely once the SSPS became operational. It is then in the nation's general interest to establish whether, or not a potential role for the SSPS may exist and, if so, to initiate investigations and discussions to elicit the issues to be resolved in allocation negotiations. The earlier this process is started, the less difficult and trying it will be to all concerned.

BODY-134   Investigations into methods of improving solar cell efficiency are extremely important to the weight and size reductions required for SSPS. The efficiency must be increased from about 14% to 18%, while at the same time reducing the thickness of the devices from about 250 to 50 μm. The overall task is expected to require 10 years. The first three or four years will be concerned primarily with theoretical and laboratory studies of potential efficiency-improved techniques and production processes.

BODY-136   "The need for reducing the cost of solar cells is a critical factor for the SSPS and has been recognized as the prime item not only for SSPS, but also for terrestrial applications of photovoltaic energy conversion techniques. Solar cells used in the space program presently cost about $80 per watt, while the SSPS requires a cost of about $0.40 per watt."

BODY-137   Cost of the SSPS is probably the most critical factor. In the case of the array blanket, it is important that additional costs beyond the cost of the solar cell itself be kept to a very low value. The interconnection of the cells and the encapsulation in a blanket must therefore be reduced from the present level of about $180 per watt (including cell costs) to a level of about $0.60 per watt.

BODY-138   "The goal of the SSPS is to produce power at high voltage in a relatively stable thermal environment (except during predictable eclipses) over a 30-year period. The cell must be extremely light in weight, yet afford protection from the space environment. The exposure of the solar blanket to the ultraviolet radiation, as well as the particulate radiation, will require protection to ensure long life."

BODY-139   The energy requirements for all the conventional processes have to be assessed to determine areas where substantial reductions in energy use will be significant. Theoretical energy-requirement limits for performing each reaction or operation can then be determined so that the potential and objectives are known. New processes and equipment will then be designed that meet these objectives.

BODY-140   The generation of high currents induce magnetic moments which can react with the natural magnetic environment and cause torques upon the SSPS or result in internal stresses caused by the interaction "self induced" local magnetic fields. The high voltage also could lead to corona formation or other ionized gas phenomena which could reduce the life of the component. The bussing of the high currents, in addition to the magnetic effects, also has an internal resistance associated with them. By judiciously sizing the current-carrying busses and using the structure for bussing when possible, the power losses can be optimized from a weight point of view.

BODY-141   "A failure mode must either be non-destructive both to the line itself and to associated systems. A monitoring system must be provided allowing sufficient time to shut down without damage."

BODY-142   "Operational experience with the prototype SSPS will be essential to permit an orderly evolution to the very-high-volume Earth-to-orbit-to-synchronous transportation system needed for an operational SSPS."

BODY-143   For an SSPS it presently appears necessary to develop advanced high-performance propulsion systems for exclusive operation in the space environment. Ion propulsion systems are the most likely choice for LEO-to-synchronous orbit transportation systems.

BODY-144   The ion propulsion system for a space tug can be designed so it will interface with payloads delivered to LEO by the space shuttle. The payloads would be transferred to a space tug which, over a period of 6 to 12 months, would follow a spiral trajectory to synchronous orbit.

BODY-145   In general, the weak link in a space assembly is the docking interface, since it has a smaller cross-sectional area than the prime construction element. Relatively large space assemblies have been analyzed in modular space station studies and found to be controllable during assembly, providing that a prescribed build-up sequence is followed and that grossly asymmetric configurations are avoided.

BODY-146   Effects at the Receiving Antenna Site

Local ecological and environmental effects at the receiving antenna site include (a) possible hazards to organisms in the receiving area due to the microwave energy received, and (b) the effects of added heat load due to microwave-to-dc conversion inefficiencies on: both the local fauna and flora (ecological effects) and the atmosphere (urban heat island effects).

BODY-147   Depending on the location, these could include snakes, lizards, rodents, and some insects. These organisms usually constitute the most significant members of the food chain including the major herbivores (rodents) and some important carnivores (snakes). If microwave energy at the anticipated levels does prove to be lethal to some organisms in the receiver area, then the results could be felt over a wider region. The receiver area could act as a biological sink attracting in-migration of organisms replacing those killed inside the area. Alternatively, or in addition, this area could act as a source of other organisms which might reproduce rapidly if their predators are reduced, and migrate out of the receiver region.

BODY-148   There is evidence that birds can be affected at levels of the order of 25-40 mW/cm2 at least at X-band, with pulsed radiation. The evidence suggests an avoidance reaction by birds. Such effects could possibly be exploited to inhibit birds from flying into the receiving antenna area. On the other hand, birds may be attracted by a possible pleasant warming sensation of microwave radiation at least in some climates. More research would have to be conducted to more clearly determine effects of microwaves on birds before deciding the best choice of system parameters for minimum interference of and by birds.

BODY-151   There is significantly higher attenuation in the Northeast and Midwest than in the Southwest and Northwest for both high and low probability of rainfall. This is due largely to the East being generally more humid and the longer beam paths through the atmosphere.

BODY-153   We have concluded that these environmental and ecological effects of heat are negligible. Thermal energy losses from the receiving antenna are very low in absolute terms, corresponding to mean heat losses in the range of 3 to 8 W/m2, depending on the size and efficiency of the receiver. More important, these energy losses are negligible when contrasted with the natural thermal energy flux occurring virtually everywhere on the surface of the Earth at any time of the year. Similarly, the loss is trivial when compared to the energy released by cities, and consequently there will be no urban heat island effect due to the presence of the receiver.

BODY-154   Examination of Table 25 convincingly supports the arguments that the mean antenna losses are low, but there are worst possible situations that might be encountered; e.g., winter when the solar insolation is lowest, and nighttime, when there is no sun at all. Furthermore, although net radiation and solar insolation are important factors in determining the heat load impinging on local flora and fauna, they are not, in fact, of primary direct importance in determining what might be called the "environmental temperature." This is due to the fact that such organisms actually intercept both incoming solar radiation from the sky and outgoing solar radiation that is reflected from the ground and other reflecting surfaces. In addition, thermal radiation — the incoming and

BODY-155   The incident solar radiation can be very large for very short periods of time, as shown in Table 26A where representative maximum values of solar insolation at the surface of the Earth are shown, together with the solar constant, which is included for comparison purposes. A high mountain top near the equator may experience insolation values as high as 1325 W/m2 for very short durations, and solar insolation on a sunny summer noon at sea level in the mid-latitudes may be as great as 700 W/m2. These, however, are maxima, and the average insolation during the day will be much less, even in summer.

BODY-156   It may be argued that to locate the receiver in a desert may be to impose an insuperable burden on a region which is ecologically fragile under the best circumstances.

BODY-158   If the soil water supply declines and insufficient water is available to satisfy the demand, the stomata close and photosynthesis ceases. At this point the temperature of the leaf may rise, since heat is no longer being lost by evaporation. If the heat being absorbed from the environment cannot be rejected as IR radiation and/or as sensible heat conducted to the air, the leaves may gradually wilt and die.

BODY-159   "The primary environmental effects of heat losses are modifications to local climate caused by changes in the sensible or latent heat transferred to the atmosphere which, in turn, cause the air to rise, and/or, change its moisture regime. Such heat may also cause modifications in rainfall patterns in the local environment. Some effects attributed to cities may also be due to the emission of condensation nuclei and to changes in the surface roughness, which also affect turbulent motions of the air."

BODY-161   Present trends in living patterns favor very high-density residential use combined with some industry and considerable commerce. These are ecologically sound in that they are well adapted to mass transit and preserve land for farming and open space. Serious consideration should be given to combining the receiving antenna with the city it is meant to serve.

BODY-162   Tidal marshes and similar coastal wetlands have been suggested as possible sites for a receiving antenna, on the grounds that these are "waste" of little economic value. However, the extremely great ecological importance of such sites effectively prohibits this or any other use of them. The existence of offshore fisheries is wholly dependent on continued unimpaired functioning of such coastal wetlands. Unlike a meadow or a field of crops, these wetlands could be severely disturbed by the disruption associated with construction of a receiving antenna, which would require deep footings, and might even entail dredging. Even minor amounts of travel for example, through, a salt marsh for the purpose of routine maintenance or inspection could have a serious cumulative negative effect. Furthermore, the presence

BODY-163   The ozone abundance is a very important factor in determining the amount and wavelength of potential damaging UV radiation which reaches the surface of the earth.

BODY-164   The chemistry of water vapor in the upper stratosphere has been studied but there is great uncertainty regarding the possible consequences of increments in water vapor on the order of 10%. Water vapor is photodissociated to form hydrogen, hydroxyl, and hydroperoxyl radicals and hydrogen peroxide molecules which will react with ozone, and molecular and atomic oxygen. The latter constituent is abundant at this level. Since some of the NOX at lower levels is produced in the mesosphere and carried downward through the region in question, it is conceivable that changes in the water vapor content will influence the natural flux of NOX to the level of the ozone layer. Consequently, the effects of shuttle flight water vapor injection in the region of 40 to

BODY-165   "The best (e.g., most conservative) approach to evaluating the effects of booster engine NO pollution in the stratosphere is to employ Johnston's arguments and see whether the shuttle vehicle NO emissions are large compared to the natural fluxes he and others have calculated."

BODY-167   "Cell efficiency, or power output, for a 50-jum (2-mil) thick cell was projected to be 19.7% and have a power output of 26.7 mW/cm2 at beginning of life and at 300°K. This is assumed to still be an accurate projection."

BODY-168   The concept that solar arrays can be manufactured inexpensively in the future was emphasized by Paul A. Berman (67): "...it seems almost inconceivable that such a simple thing as a solar array substrate with printed circuit interconnections and wiring upon which cells are mounted in some"

BODY-171   "Three basic phases lead to an operational SSPS program; they are: Technology Development/Verification, Prototype System Development, and Operational System Development."

BODY-172   The combination of advanced single stage to orbit transportation system elements and assembly operations at less than synchronous altitudes provide optimism that a $50/lb goal, for transportation costs associated with the delivery of the operational SSPS, is potentially achievable by the 1990's.

BODY-173   Cost estimates for the microwave generators can be based on the substantial experience with state-of-the-art microwave devices. The cost estimates on the transmitting antenna are less firm, because detailed design of the antenna has not yet been accomplished. Space structures have a high cost when measured in terms of dollars per pound of structure compared to the cost of commercially produced structures. However, most of the structure of the transmitting antenna will be highly repetitive and will be produced by mass production techniques. Because of the need for light-weight structures, the largest part of the cost will be in light-weight, high-performance materials.

BODY-176   The composite figure is more than a simple sum. The figure for 0.25 which represents the overall cost of the system must be greater than the sum of individual figures in the 0.25 column — how much greater depends on the degree of independence of the individual events and the nature of the individual distribution (i.e., the curve between and beyond the points 0.25, 0.50, and 0.75). The composite figures* are based on very simple assumptions:

(1) The events are completely independent; and

(2) The distribution is nearly rectangular.

BODY-177   The costs and benefits associated with the project are evaluated to determine the economic feasibility of an investment of this type. Such an evaluation asks the question whether the expenditure of funds on this project will produce a product which will earn the total economy a satisfactory rate of return and thereby be considered as a worthy undertaking. This analysis must take into account the social and environmental side effects of both the old and the new technology.

BODY-178 Cost/Benefit analysis has been a reliable analytical tool for policy-makers in the process of making research developmental and operational investment decisions. At the same time, its limitations are well known, perhaps best to those who have relied on it most. This does not suggest that the use of cost/benefit analysis on decisions of a long-term nature transcending national importance should be discouraged, but only that sober judgment of its promise in dealing with such decisions be applied.

BODY-179 The order of magnitude of the investment required to do the research and development for the SSPS will be comparable to the actual costs of developing nuclear power or any other large undertaking with a similar potentially significant impact on a national and eventually worldwide basis.

BODY-180 "Evaluations of power supply increments per se are neither useful nor relevant for planning; rather, evaluations must be made of existing systems with and without the new supply increments. This holds for conventional systems as well as mixtures of new and old technology."

BODY-181 "Lessening of environmental impacts will play a greater and greater part in the selection and management of energy sources and energy-producing technology. These problems lead in three directions for relief: (1) ways to lessen the impact of present and proven energy technology on the environment; (2) palatable public policy which might lessen the demand for energy; and (3) new and more favorable technology."

BODY-182 The competitive analysis will be more variable since the amount of information known and available about any power generation differs from scheme to scheme. This difference in information is characteristic of elements in any portfolio and is part of the reason why the methodology of evaluation should be based on a portfolio concept. For example, the direct cost of proven and available technology will be readily available, whereas the possible cost of a fusion system may be essentially unknown.

BODY-183 New technologies such as SSPS, fusion reactors, arid breeder reactors, may have near-zero operating costs. Thus the simple rule-of-thumb "dollars per kilowatt" becomes less significant, and it is increasingly desirable to think about mills per kilowatt-hour.

BODY-184 In the long run, however, it is necessary that the load-duration characteristics of demand be taken into consideration. Only 40% of the kilowatt capacity installed in a system can be operated in the 80-85% range; for the rest, the load simply is not there.

BODY-185 "It has been forecast that between now and 2001 the United States will consume more energy than it has in its entire history, and that by 2001 the annual U.S. demand for energy in all forms will double and the annual world-wide demand will probably triple. These projected increases will tax man's ability to discover, extract, and refine fuels in the huge volumes necessary to ship them safely and to dispose of waste products with minimum harm to himself and his environment."

BODY-186 "Preliminary estimates indicate that this energy tax against the SSPS can be met by less than one year of operation."

BODY-187 An analysis of the effects an investment might have on the structure of the economy can best be accomplished by using an input-output (I/O) model of the economy to test the industry effect. Basically, this model answers the following question: What will the effects on other industries be? To illustrate how this question can be *answered the automobile industry can serve as an example. The I/O model specifies the production technology of the automobile industry. It will show that automobiles require certain inputs from the steel industry, the plastics industry, the rubber industry and so much labor. These amounts are specified. Therefore, an increase in demand for cars will create a derived demand for products from other industries.*

BODY-189   *An important input into the above analysis is the rate at which final demand will change. A very rough estimate could be inserted here, possibly by asking the relevant people; or an alternative course would be to simulate the entire systems for different rates of adjustment. This path would provide useful information about the effects on different industries from a shift to the SSPS as the source of power.*

BODY-190   *The flight control performance evaluation indicated that the pointing accuracy of the SSPS fell well within the ± 1-deg limit specified by the baseline requirements for the pitch, roll, and yaw axes. In addition, the system's response time and percent overshoot were found to be acceptable for control about all three axes.*

BODY-191   *These efforts should ultimately be focused on the identification of the minimum-weight spacecraft system and structure having acceptable structural stiffness and pointing capabilities.*

BODY-192   *Successful demonstration of high-efficiency rectification of microwaves to dc is expected to show that the SSPS will be capable of generating power on Earth with an efficiency which has not yet been equalled by any known power generation method.*

BODY-193   *Thus, we feel that engineering, research, and development, rather than any scientific breakthrough, represent the keys to their solution.*

BODY-194   *"The SSPS can be designed to accommodate a wide range of microwave power flux densities to meet internationally accepted standards of microwave exposure. The transmitting antenna size, the shape of the microwave power distribution across the antenna, and the total power transmitted will determine the level of microwave power flux densities in the beam reaching the Earth."*

BODY-195   *An overall SSPS-related biological effects program should be carried out and closely integrated with the research plans for a national biological effects program being developed by the Office of Telecommunications Policy.*

BODY-196   *Fulfillment of these goals will require a better definition of the SSPS prototype's desirable performance, size, and estimated cost. Once these have been broadly established, a finer grained structure of technology and cost goals can be folded into present technology development and verification plans.*

BODY-203   *"The aeronautical and space activities of the United States shall be conducted so as to contribute . . . to the expansion of human knowledge of phenomena in the atmosphere and space. The Administration shall provide for the widest practicable and appropriate dissemination of information concerning its activities and the results thereof." —NATIONAL AERONAUTICS AND SPACE ACT OF 1958*

FEASIBILITY STUDY OF
A SATELLITE SOLAR POWER STATION

by Peter E. Glaser, Owen E. Maynard,
John Mackovciak, Jr., and Eugene L. Ralph

Prepared by
ARTHUR D. LITTLE, INC.
Cambridge, Mass. 02140
for Lewis Research Center

NATIONAL AERONAUTICS AND SPACE ADMINISTRATION • WASHINGTON, D. C. • FEBRUARY 1974

ERRATA

NASA Contractor Report CR-2357

FEASIBILITY STUDY OF A SATELLITE SOLAR POWER STATION

by Peter E. Glaser, Owen E. Maynard, John Mackovciak, Jr.,
and Eugene L. Ralph
February 1974

Page 8, Table 1, line 1: The value for Gravity Gradient should be 12.1×10^{4} newton-m sec.

Page 8, Table 1, line 8: The value in parentheses following RCS should be I_{sp} 800 sec

REVISION OF SECTION

On

STRATOSPHERIC POLLUTION WITH
SHUTTLE VEHICLE EXHAUST PRODUCTS

Pages 149-153

Of

NASA CR-2357

FEASIBILITY STUDY OF A
SATELLITE SOLAR POWER STATION

By

Peter E. Glaser
Owen E. Maynard
John Mackovciak, Jr.

The enclosed revision of pages 149-153 of
NASA CR-2357 contains several corrections
of the original material and reflects the
current Space Shuttle performance characteristics.

1. Report No. NASA CR-2357	2. Government Accession No.	3. Recipient's Catalog No.
4. Title and Subtitle FEASIBILITY STUDY OF A SATELLITE SOLAR POWER STATION		5. Report Date February 1974
		6. Performing Organization Code
7. Author(s) Peter E. Glaser, Owen E. Maynard, John Mackovciak, Jr., and Eugene L. Ralph (see following page for affiliation)		8. Performing Organization Report No. ADL-C-74830
		10. Work Unit No.
9. Performing Organization Name and Address Arthur D. Little, Inc. 20 Acorn Park Cambridge, Massachusetts 02140		11. Contract or Grant No. NAS 3-16804
		13. Type of Report and Period Covered Contractor Report
12. Sponsoring Agency Name and Address National Aeronautics and Space Administration Washington, D.C. 20546		14. Sponsoring Agency Code

15. Supplementary Notes

Final Report. Project Manager, Ronald L. Thomas, Power Systems Division, NASA Lewis Research Center, Cleveland, Ohio

16. Abstract

A feasibility study of a satellite solar power station (SSPS) was conducted to (1) explore how an SSPS could be "flown" and controlled in orbit; (2) determine the techniques needed to avoid radio frequency interference (RFI); and (3) determine the key environmental, technological, and economic issues involved. Structural and dynamic analyses of the SSPS structure were performed, and deflections and internal member loads were determined. Desirable material characteristics were assessed and technology developments identified. Flight control performance of the SSPS baseline design was evaluated and parametric sizing studies were performed. The study of RFI avoidance techniques covered (1) optimization of the microwave transmission system; (2) device design and expected RFI; and (3) SSPS RFI effects. The identification of key issues involved (1) microwave generation, transmission, and rectification and solar energy conversion; (2) environmental-ecological impact and biological effects; and (3) economic issues, i.e., costs and benefits associated with the SSPS. The feasibility of the SSPS based on the parameters of the study was established.

17. Key Words (Suggested by Author(s)) Solar cells; Satellite power system; Microwave power; Solar terrestrial power	18. Distribution Statement Unclassified - unlimited		
19. Security Classif. (of this report) Unclassified	20. Security Classif. (of this page) Unclassified	21. No. of Pages 199	22. Price* Domestic, $5.25 Foreign, $7.75

* For sale by the National Technical Information Service, Springfield, Virginia 22151

Authors: Peter E. Glaser, Arthur D. Little, Inc., Cambridge, Massachusetts; Owen
E. Maynard, Raytheon Company, Sudbury, Massachusetts; John Mockovciak, Jr.,
Grumman Aerospace Corporation, Bethpage, New York; and Eugene L. Ralph, Spectro-
lab (Textron Inc.), Sylmar, California

TABLE OF CONTENTS

LIST OF FIGURES

LIST OF TABLES

SUMMARY

This is the Final Report of a feasibility study of a satellite solar power station (SSPS) carried out by Arthur D. Little, Inc., Grumman Aerospace Corporation, Raytheon Company, Spectrolab, a division of Textron Inc., for the National Aeronautics and Space Administration under Contract NAS 3-16804. The primary objectives of this study were (1) to explore how an SSPS could be "flown" and controlled in orbit, (2) to determine the techniques which would be required to avoid radio frequency interference with other users of the electromagnetic spectrum, and (3) to determine the key environmental and economic issues which would have to be assessed.

Structure and Control Techniques

Structural and dynamic analyses of the SSPS structure were performed to provide elastic characteristics (natural frequencies, generalized masses, and mode shapes) of the structure for use in an analytical investigation of the elastic coupling between the SSPS attitude control system and the spacecraft's structural modes. Deflections and internal member loads resulting from the various flight loading conditions were determined to verify structural integrity.

Desirable material characteristics were assessed and technology developments identified to provide inputs leading to the design of structure and attitude control systems for the very large-area, light-weight space structures represented by the SSPS.

The flight control performance of the SSPS baseline design was evaluated and parametric sizing studies performed to determine the influence of structural flexibility upon attitude control system performance.

RFI Avoidance Techniques

The study of RFI avoidance techniques included three principal areas: (1) optimization of the microwave transmission system; (2) device design and expected RFI; and (3) effects of SSPS RFI on other users.

System Optimization. — To optimize the microwave transmission system, a model and a set of assumptions were first defined. The model included data on orbital and ground location, ground power transmission, device characteristics, phase-front control, efficiencies, RF environment, attenuation, frequencies, users, and equipment.

Device Design. — The Amplitron, a very efficient microwave generator, was evaluated from the viewpoint of its design versus its operating frequency for the SSPS concept. The choice of 3.3 GHz as the fundamental frequency for the SSPS was based on a set of assumptions for filter design and recognition of existing allocated radio astronomy and fixed satellite space-to-Earth bands.

Effects of SSPS RFI on Other Users. — This phase of the study concentrated on (1) the transmitting antenna and nature of the transmitted beam; (2) the receiving antenna; and (3) noise

1

emanating from the SSPS. The analysis showed that the approach required for Amplitron design and filtering techniques would minimize RFI with other users, and hence national and international agreement on frequency allocation for the SSPS would be achievable.

Identification of Key Issues

Technological Issues. —

a. Microwave Generation, Transmission and Rectification

The microwave portion of the electromagnetic spectrum has been selected as the most useful for SSPS power generation, transmission, and rectification. From a device point of view, the Amplitron appears most promising because of its unique performance characteristics.

b. Solar Energy Conversion

Design approaches for the solar collector, solar cell blankets, and power collection and distribution methods were evolved to meet the requirements of the structure and control technique analyses.

Environmental Issues. —

a. Environmental/Ecological Impact

The environmental and ecological impacts of the SSPS were explored, with attention focussed on the environmental and ecological impact at the receiving antenna, and the possible contamination of the stratosphere by the space transportation system. Waste heat released at the receiving antenna does not constitute a significant thermal effect on the atmosphere, and, with RF shielding incorporated below the rectifying elements, the receiving antenna operation can be compatible with other land uses.

b. Biological Effects

There exist conflicting interpretations of the effects of microwave exposure throughout the scientific community. Because of the lack of internationally accepted standards, based on experimental data, to place a specific and allowable level on microwave exposure, the SSPS will have to be designed to accommodate a wide range of microwave power flux densities.

Economic Issues. — Three of the key issues that will have to be addressed when making an economic comparison of the SSPS with other means of generating power and the methodology to deal with these issues include:

1. The *costs and benefits* associated with the SSPS which have to be evaluated to determine the economic feasibility of an investment of this type.

2

2. The *macro-economic interindustry effects* produced by the SSPS which have to be examined to analyze the effects such an investment might have on the *structure of the economy* as a whole.

3. The *consumption effects* created by the SSPS which will be reflected in both the cost/benefit analysis and the analysis of macro-economic interindustry effects.

GLOSSARY OF SYMBOLS

Symbol	Description		
$F_{c_{x,y,z}}$	Attitude control force along the x,y,z axes, respectively	kg	slugs
I_{xx}, I_{yy}, I_{zz}	Principal moment of inertia about the x,y,z axes respectively	kg-m^2	slug-ft^2
$K_{A_{x,y,z}}$	Attitude control system gain		
$K_{d_{x,y,z}}$	Portion of external disturbance torque that is proportional to attitude angle errors acting about the x,y,z axes, respectively	kg-m/rad	ft-lb/rad
$K_{R_{x,y,z}}$	Rate sensor gain	rad/(rad/sec)	
l_x, l_y	Spacecraft length along the x and y axes, respectively	m	ft
q_i	Generalized displacement of the ith mode	m	ft
$T_{d_{x,y,z}}$	Portion of external disturbance torque that is constant and acting about the x,y,z axes, respectively	kg-m	ft-lb
$T_{R_{x,y,z}}$	Control system response time for the x,y,z axes, respectively	sec	
δ	Rigid body damping ratio		
$\theta_{x,y,z}$	Spacecraft rigid body rotational attitude about the x,y,z axes, respectively	rad	
$\theta_{c_{x,y,z}}$	Rotational attitude commands about the x,y,z axes, respectively	rad	
$\Delta\theta_x$	Rotational attitude errors about the x,y,z axes, respectively	rad	

σ_i	Normalized slope of the ith mode at the right end	rad/m	rad/ft
ϕ_i	Normalized modal deflection of the ith mode at the right end	m/m	ft/ft
ω_n	Undamped natural frequency	rad/sec	
ω_1	Natural frequency of 1st anti-symmetric mode	rad/sec	

INTRODUCTION

Solar energy is being seriously considered as an alternative energy source for a wide range of applications not only as a result of technological advances, but in response to a variety of economic, environmental, and social forces. As limitations on conventional energy sources and the environmental consequences of energy production become more apparent, solar energy stands out as an inexhaustible alternate energy source if it can be harnessed within economic, environmental, and social constraints.

Recently the potential of solar energy to meet future needs has been re-examined (1). Today opportunities for harnessing solar energy, both over the long and short term, are being investigated by government and industry.

The magnitude of solar energy theoretically available is far in excess of future needs. Although the sun radiates vast quantities of energy, they reach the Earth in a very dilute form. Thus, any attempts to harness solar energy on a significant scale will require devices which occupy a large area as well as locations that receive a copious supply of sunlight. These requirements restrict Earth-based solar energy conversion devices for producing power to a few favorable geographical locations. Even for these locations energy storage must be provided to compensate for the day-night cycle and cloudy weather.

One way to harness solar energy effectively would be to move the solar-energy conversion devices off the surface of the Earth and place them in orbit away from the Earth's active environment and influence and resulting erosive forces (2). The most favorable orbit from the power density point of view would be one around the sun, but a synchronized orbit around the Earth could be used where solar energy is available nearly 24 hours of every day.

In the five years since the concept of a satellite solar power station (SSPS) was first presented as an alternative energy production method (3), the energy crisis experienced in the technologically advanced countries has intensified because of increasing energy use and demands for a clean environment.

An assessment of the feasibility of the SSPS concept has shown that it is worthy of consideration as an alternative energy production method (4-9). Its development can be realized by building on scientific realities, on an existing industrial capacity for mass production, and on demonstrated technological achievement (10).

BASELINE DESIGN

Principles of a Satellite Solar Power Station

Figure 1 shows the design concept for an SSPS. Two symmetrically arranged solar collectors convert solar energy directly to electricity by the photovoltaic process while the satellite is maintained in synchronous orbit around the Earth. The electricity is fed to microwave generators incorporated in a transmitting antenna located between the two solar collectors. The antenna directs the microwave beam to a receiving antenna on Earth where the microwave energy is efficiently and safely converted back to electricity.

FIGURE 1   DESIGN CONCEPT FOR A SATELLITE SOLAR POWER STATION

An SSPS can be designed to generate electrical power on Earth at any specific level. However, for a power output ranging from about 3,000 to 15,000 MW, the orbiting portion of the SSPS exhibits the best power-to-weight characteristics. Additional solar collectors and antennas could be added to establish an SSPS system at a desired orbital location. Power can be delivered to most desired geographic locations with the receiving antenna placed either on land or on platforms over water near major load centers, and tied into a power transmission grid. The status of technology and the advances which will be required to achieve effective operation for an SSPS are described below.

Location of Orbit. — The preferred locations for the SSPS are the Earth's equatorial synchronous orbit stable nodes which occur near the minor axes, at a longitude of about 123° West

6

and 57° East. The minor axes are stable node points and the major axes unstable. The SSPS would be positioned so its solar collectors always face the sun, while the antenna directs a microwave beam to a receiving antenna on Earth. The microwave beam would permit all-weather transmission so that full use could be made of the nearly 24 hours of available solar energy. In an equatorial, synchronous orbit, the satellite can be maintained stationary with respect to any desired location on Earth.

There are three major influences on the SSPS which would cause it to drift from its nominal orbital location (Figure 2):

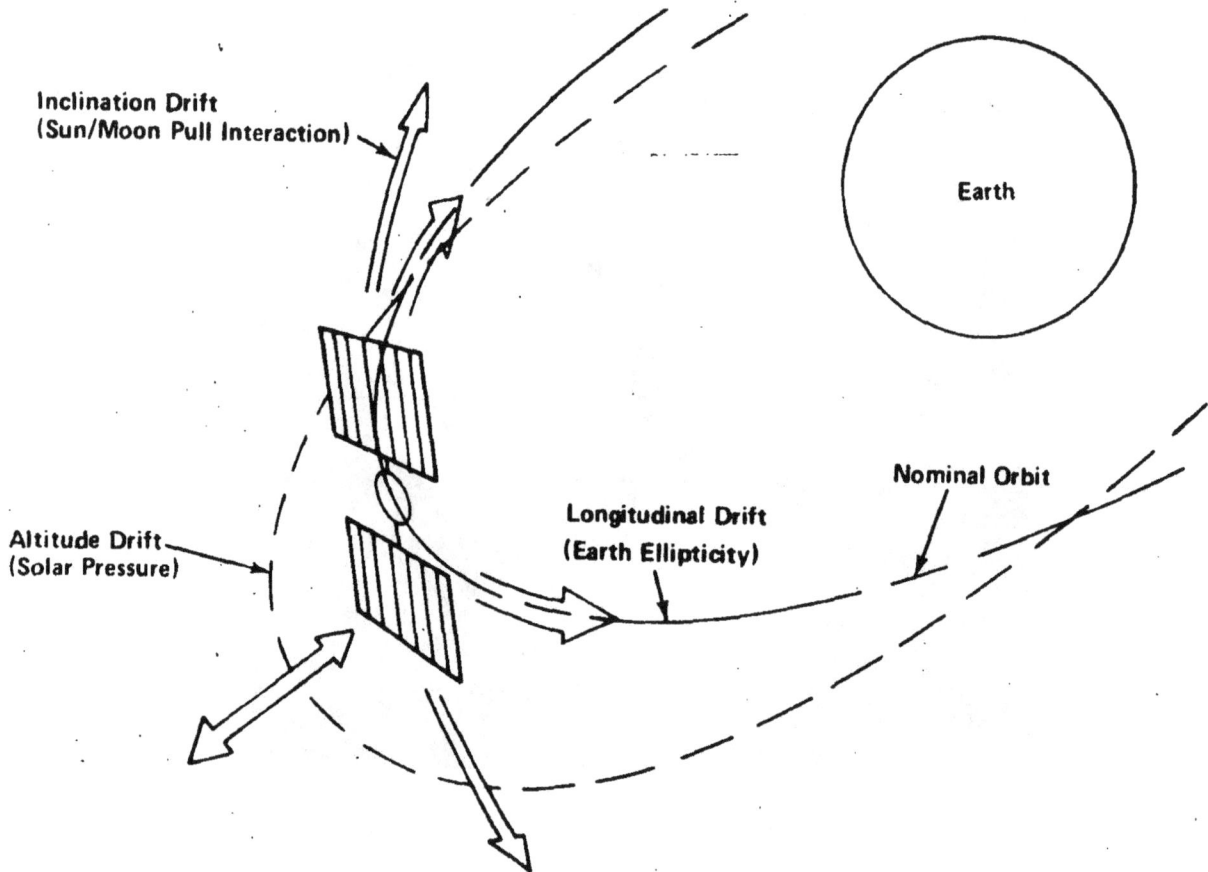

FIGURE 2. – NOMINAL ORBIT PERTURBATIONS

1. The ellipticity of the Earth causes the SSPS to seek out the Earth's minor axes;

2. The interaction of the gravitational effects of the sun and the moon would cause the orbit to regress so that its inclination would change with respect to the equator; and

3. Solar pressure would distort the orbit from circular to elliptical and back again over a one-year period. In addition, there would be an effective altitude change which

7

would increase the orbital period and then restore it to nominal over the same elapsed time.

A summary of the flight control requirements is presented in Table 1.

TABLE 1

FLIGHT CONTROL REQUIREMENTS

Gravity Gradient	1.67×10^9 newtons m sec
Solar Pressure	$T_y = 5500$ newton-m
	$T_x = 136$ newton-m
Electromagnetic Field Interactions	$< 10^{-5}$ newton-m
Rotary Joint Friction	216 newton-m
Microwave Transmission Recoil Pressure	11 newtons
Aerodynamic	22×10^{-6} newton
RCS (Ion Thruster, ISP 8000 sec) Propellant Weight to Control SSPS to within 1 Deg	43.6 kg/day
	15,500 kg/year

An SSPS in synchronous equatorial orbit would pass through the Earth's shadow around the time of equinoxes, at which time it would be eclipsed for a maximum of 72 minutes a day (near midnight at the SSPS longitude). This orbit provides a 6- to 15-fold conversion advantage over solar-energy conversion on Earth. A comparison of the maximum allowable costs of photovoltaic energy-conversion devices indicates that for a terrestrial solar-power application these devices are competitive with other energy-conversion methods if they cost about $2.30 per square meter. Because of the favorable conditions for energy conversion that exist in space, these devices are competitive if they cost about $45 per square meter in an Earth-orbit application (11).

Solar Energy Conversion. — The photovoltaic conversion of solar energy into electricity is ideally suited to the purposes of an SSPS. In contrast to any process based on thermodynamic energy conversion, there are no moving parts, fluid does not circulate, no material is consumed, and a photovoltaic solar cell can operate for long periods without maintenance. There has been a substantial development in photovoltaic energy conversion since the first laboratory demonstration of the silicon solar cell in 1953. Today, such cells are a necessary part of the power supply system of nearly every unmanned spacecraft, and considerable experience has been accumulated to achieve long-term and reliable operations under the conditions existing in space. Thus, the "Skylab" spacecraft relies on silicon solar cells to provide about 25 kW of power. As a result of many years of operational experience, a substantial technological base exists on which further developments can be based (12). These developments are directed towards increasing the efficiency of solar cells, reducing their weight and cost, and maintaining their operation over extended periods.

8

a. Efficiency Increase

The maximum theoretical efficiency of a silicon solar cell is about 22%. The most widely used single-crystal silicon solar cells can routinely reach an efficiency of 11% and efficiencies of up to 16% have been reported (13). Development programs to increase the efficiency of silicon solar cells up to 20% have been outlined (14).

The silicon solar cell which is produced from single-crystal silicon is typically arranged with the P-N junctions positioned *horizontally*. More recently, *vertically* illuminated multijunction silicon solar cells have been investigated (15). These have the potential to reach higher efficiencies and to be more resistant to solar radiation damage.

Solar cells made from single-crystal gallium arsenide exhibit an efficiency of about 14% with a theoretical limit of about 26%. Recently, a modified gallium arsenide solar cell was reported to have reached an efficiency of 18% (16). This substantial increase in efficiency is particularly significant, because these cells can operate at higher temperatures than silicon solar cells, are more radiation-resistant and can be prepared in thicknesses about one-tenth that of a silicon solar cell.

Several other materials may be suitable for photovoltaic solar energy conversion. Among these are various combinations of inorganic semiconductors which have only partially been investigated. Organic semiconductors which exhibit the photovoltaic effect and which do not have known boundaries to the theoretical efficiency also remain to be explored, so that their potential for photovoltaic solar-energy conversion can be established (17).

b. Weight

Single-crystal silicon solar cells are presently 500 to 1000 microns thick, although their thickness could be reduced to about 50 microns without compromising efficiency. However, gallium arsenide cells need be only a few microns thick.

The individual solar cells have to be assembled to form the solar collector. The weight of a solar cell array can be reduced by assembling the solar cells in a blanket between thin plastic films, with electrical interconnections between individual cells obtained by vacuum-depositing metal alloy contact materials. The collector weight can be further reduced when solar energy concentrating mirrors arranged to form flatplate channels are used so that a smaller area of solar cells is required for the same electrical power output (see Figure 3). The weight and cost of a given area of a reflecting mirror used to concentrate solar energy are considerably less than those for the same area of solar cells.

Suitable coatings on mirrors to reflect only the component of the solar spectrum most useful for photovoltaic conversion can reduce heating of the solar cells and thus increase efficiency. There is a balance between concentrating the solar radiation per unit area of the cell, which may lead to a rise in temperature and a consequent decrease in solar cell efficiency, and the desire to maximize the collection of solar energy. An array configuration that includes mirrors with a concentration factor of about 2 has been chosen for the solar collector.

9

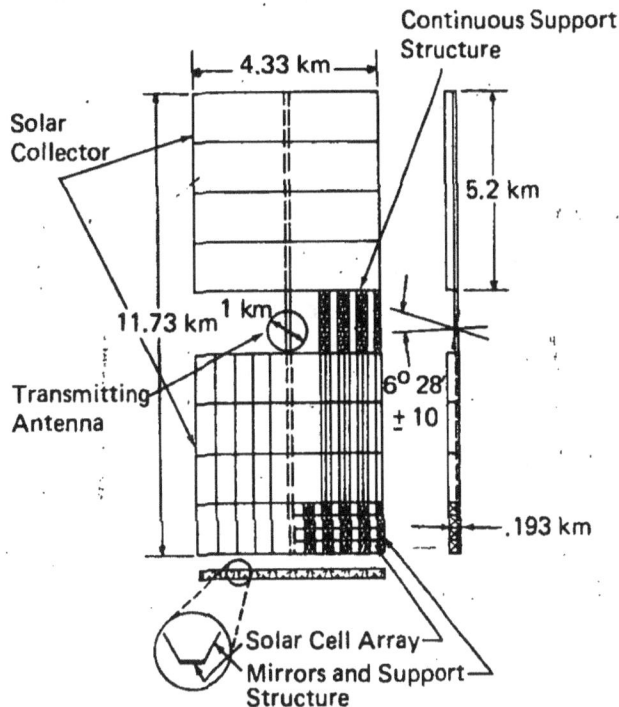

FIGURE 3. – SSPS DIMENSIONS

Since 1965, the solar cell weight has decreased substantially. With the development of blanket-type construction for solar cell arrays, this weight is projected to drop to 4 kg/kW by 1975. The use of solar energy-concentrating mirrors can reduce the solar collector weight to about 1 kg/kW if 100-micron thick silicon solar cells are used, and to even less if gallium arsenide solar cells are used.

c. Electrical Interconnections

Because solar cell arrays with large areas will be required (e.g., one of the two solar arrays for the SSPS is about 5 km by 5 km), the satellite structures must be designed to combine mechanical and power distribution network functions to achieve high-voltage DC output. In the very large arrays, solar cells can be connected in series to produce any voltage desired. With existing solar cells, a series string can be assembled to build up the voltage to 50 kV or more.

Development of the vertically illuminated, multijunction solar cell could product solar cells of high-voltage output. In such a solar cell, there may be a thousand junctions in series for each 1-cm wide cell. Thus, each cell may put out several hundred volts instead of the 0.5 volt from present solar cells. This type of cell will make it easier to build up a high voltage with a small number of cells, and thus allow most of the circuit to be in parallel and to be less susceptible to losses from open circuits. This arrangement also would preclude the loss of a total string of solar cells because of the loss of any one link.

10

The power bus interconnecting the major segments of the solar cell arrays, which will have to carry several hundred thousand amperes, must be designed to minimize magnetic field interactions. This can be done by suitable arrangement of the power distribution circuits. High-voltage switching circuits will have to be developed to control sections of the solar cell array for maintenance and operational purposes, and to protect the solar cell arrays when they enter and leave the Earth's shadow.

The system also must provide the capability of switching off all power by open-circuiting solar cells instantaneously in the event of system malfunction. Because the SSPS system provides no energy storage, it will be safer than conversion systems that rely on thermodynamic power.

d. Effects of the Space Environment

The state of the art of solar cells is now at a level where lifetimes of 10 years are achievable. For example, the effective life expectancy of the Intelsat IV satellite is eight years. But the operations of the solar cell arrays will be influenced by the space environment.

One influencing factor in space will be solar radiation. Solar radiation damage will cause a logarithmic decay of solar-cell effectiveness. However, improvements in radiation-resistant solar cells are expected to result in a 30-year minimum operational lifetime for the SSPS, after which normal SSPS effectiveness can be restored by adding a small area of new solar cells. Thus, there will be no absolute time limit on effective SSPS operation.

Another aspect of the space environment that will influence SSPS operations is the impact of micrometeoroids. In synchronous orbit, the SSPS is expected to suffer a 1% loss of solar cells, based on the probability of damage-causing impacts by micrometeoroids during a 30-year period.

The benign nature of the space environment and the absence of significant gravitational forces, however, permit the design of solar collector arrays which have a minimum material mass. In addition, their performance would be much more predictable than that of an Earth-based solar energy conversion device because of the absence of the vagaries of the Earth environment.

Microwave Power Generation, Transmission, and Rectification. — The power generated by the SSPS in synchronous orbit must be transmitted to a receiving antenna on the surface of the Earth and then rectified. The power must be in a form suitable for efficient transmission in large amounts across long distances with minimum losses and without affecting the ionosphere and atmosphere. The power flux densities received on Earth must also be at levels which will not produce undesirable environmental or biological effects. Finally, the power must be in a form that can be converted, transmitted, and rectified with very high efficiency by known devices.

All these conditions can best be met by a beam link in the microwave part of the spectrum. In this part of the spectrum a desirable frequency can be selected, e.g., about 3.3 GHz, and induced radio frequency interference limited so that an appropriate internationally agreed upon frequency could be assigned to an SSPS.

11

Fortunately, man has considerable experience in high-power microwave generation, transmission, and rectification. As early as 1963, Brown (18) demonstrated that large amounts of power could be transmitted by microwaves. The efficiency of microwave power transmission will be high when the transmitting antenna in the SSPS and the receiving antenna on Earth are large. The dimensions of the transmitting antenna and the receiving antenna on Earth are governed by the distance between them and the choice of wavelength (19).

The size of the transmitting antenna is also influenced by the inefficiency of the microwave generators due to the area required for passive radiators to reject waste heat to space and the structural considerations as determined by the arrangement of the individual microwave generators. The size and weight of the transmitting antenna will be reduced as the average microwave power flux density on the ground is reduced by increasing the size of the receiving antenna and as higher-frequency microwave transmission is used. The size of the receiving antenna will be influenced by the choice of the acceptable microwave power flux density, the illumination pattern across the antenna face, and the minimum microwave power flux density required for efficient microwave rectification.

a. Microwave Attenuation (20)

Ionospheric attenuation of microwaves is low (less than 0.1%) for wavelengths between 3 and 30 cm and for the microwave power flux densities occurring within the beam. Tropospheric attenuation is low for wavelengths near and above 10 cm, but attenuation will increase as wavelengths are reduced. Moderate rainfall attenuates microwaves approximately 10% at a wavelength of 3 cm and 3% at a wavelength of 10 cm at a nadir angle of 60 deg. The efficiency of transmission through the atmosphere in temperate latitudes, including some rain (2 mm/hr), is approximately 98% and decreases to 94% for moderate (33 mm/hr) rainfall, depending on location (see Figure 4).

b. Microwave Transmission System Efficiency

The efficiency of the microwave power transmission system is a product of the efficiency of dc-to-microwave power conversion, the efficiency with which the microwave power is transmitted to the receiving antenna by the microwave beam link, and the efficiency with which the microwave beam is controlled, pointed toward, and intercepted at the receiving antenna, and there rectified to dc. This overall transmission efficiency can also be measured experimentally as a ratio of the dc power output at the receiving antenna to the dc power input at the transmitting antenna. Table 2 indicates the efficiencies that have been demonstrated in the three major functional categories of the microwave transmission system and the projected efficiencies which should be achievable with further development (21).

Including microwave attenuation, the overall efficiency of microwave transmission from dc in the SSPS to dc on the ground is projected to be about 70%.

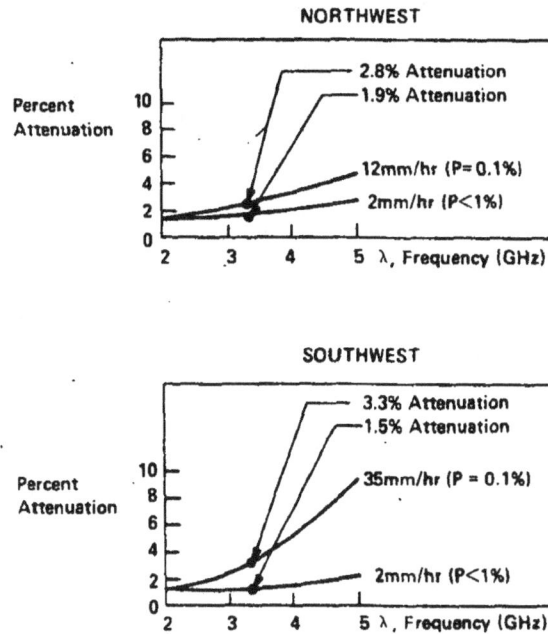

FIGURE 4. – ATMOSPHERIC ATTENUATION OF MICROWAVES
IN TWO UNITED STATES LOCATIONS

c. Microwave Generation

The microwave generator design is based on the principle of a crossed-field device which has the potential to achieve a high reliability and a very long life (22). A pure-metal, self-starting, secondary emitting, cold cathode is employed in a non-reentrant circuit, matched to an input and output circuit so as to provide a broadband gain device. The device is designed to be capable of automatically self-regulating its power output. The use of samarium-cobalt permanent magnet material leads to substantial weight reduction compared to previously available magnet material. The vacuum in space obviates the glass envelope required on Earth. The cathode and anode of the microwave generator are designed to reject waste heat with passive extended-surface radiators which radiate to space. The output of an individual microwave generator weighing a fraction of a kilogram per kilowatt can range from 2 to 5 kW.

TABLE 2

MICROWAVE POWER TRANSMISSION EFFICIENCIES

	Efficiency Presently Demonstrated[a]	Efficiency Expected with Present Technology[a]	Efficiency Expected with Additional Development[a]
Microwave Power Generation Efficiency	76.7[b]	85.0	90.0
Transmission Efficiency from Output of Generator-to-Collector Aperture	94.0	94.0	95.0
Collection and Rectification Efficiency (Rectenna)	64.0	75.0	90.0
Transmission, Collection, and Rectification Efficiency	60.2	70.5	85.0
Overall Efficiency	26.5[c]	60.0	77.0

a. Frequency of 2450 MHz (12.2-cm wavelength).

b. This efficiency was demonstrated at 3000 MHz and a power level of 300 kw CW.

c. This value could be immediately increased to 45% if an efficiency generator were available at the same power level at which the efficiency of 60.2% was obtained.

The quantity of 1 to 2 million tubes that would be needed for each SSPS is large enough to warrant large-scale, highly efficient mass production. There is substantial production experience on magnetrons, similar in many respects to the Amplitron device projected for use in the SSPS.

d. Microwave Transmission

A series of microwave generators will be combined in a subarray (e.g., about 5 m by 5 m) which forms part of the antenna. Each subarray must be provided with an automatic phasing system so the individual antenna radiating elements will be in phase. These subarrays will radiate through a slotted waveguide and form a phased-array transmitting antenna about 1 km in diameter to obtain a microwave beam of a desired distribution. The distribution can be designed to range from uniform to near gaussian.

For this 1-km diameter antenna, the diameter of the receiving antenna on Earth would have to be about 7 km for gaussian distribution in the beam within which 90% of the transmitted energy is intercepted. The use of such a large receiving antenna area would reduce the microwave power flux density on the Earth to a value low enough so that the flux density of the edges of the receiving

14

antenna would be substantially less than the continuous microwave exposure standard presently accepted in the United States (i.e., 10mW/cm^2). Within several kilometers beyond the receiving antenna, the microwave density levels drop to less than $1\ \mu\text{W/cm}^2$(23). (See Figure 5.) The data in Figure 5 are for a total power of 1.17×10^7 kW for a single SSPS whereas the SSPS baseline design is for about half that value. The curves, however, indicate the significant decay rate with increased radius.

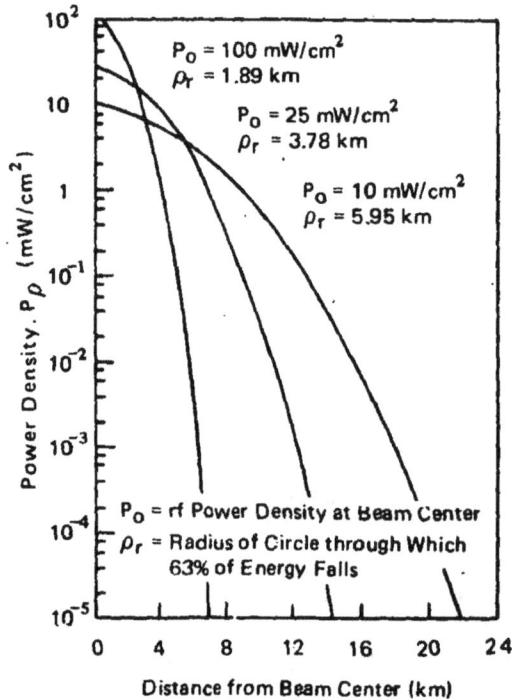

Note: Total Power held constant at 1.17×10^7 kw

FIGURE 5. — HYPOTHETICAL DISTRIBUTIONS OF MICROWAVE
POWER DENSITY FROM THE BEAM CENTER
(Ideal Gaussian)

To achieve the desired high efficiency for microwave transmission, the phased-array antenna will be pointed by electronic phase shifters (24). Proper phase setting for each subarray must be established to form and maintain the desired phase front. Deviations can be detected and appropriate phase shifts made to minimize microwave beam scattering. A master phase control in the antenna will have to be developed if the microwaves are to be transmitted efficiently and the microwave beam always directed toward the receiving antenna (25). The master phase front control system can be designed to compensate for the tolerance and position differences between the subarrays by sensing the phase of a pilot signal beamed from the center of the area occupied by the receiving antenna to control the phase of the microwaves transmitted by each subarray. The pilot signal will be of a substantially different frequency than that of the microwave power beam, so wave filters could be used to separate them.

15

Precision pointing of the receiving antenna is not necessary to the operation of the SSPS and inhomogeneities in the propagation path are not significant. Any deviation of the microwave beam beyond allowable limits would preclude acquisition of the pilot signal. Without the signal, the coherence of the microwave beam would be lost, the energy dissipated, and the beam spread out so the microwave power density would approach communication signal levels. This phase-control approach would assure that the beam could not be directed either accidentally or deliberately towards any other location but the receiving antenna. This inherent fail-safe feature of the microwave transmission system is backed up by the operation of the switching devices, which would open-circuit the solar cell arrays to interrupt the power supply to the microwave generators.

e. Microwave Rectification

The receiving antenna is designed to intercept, collect, and rectify the microwave beam into dc which can then be fed into a high-voltage dc transmission network or converted into 60-Hz ac. Half-wave dipoles distributed throughout the receiving antenna capture the microwave power and deliver it to solid-state microwave rectifiers (26). Schottky barrier diodes have already been demonstrated to have a 80% microwave rectification efficiency at 5W of power output. With improved circuits and diodes, a rectification efficiency of about 90% will be achievable.

The diodes combined with circuit elements which act as half-wave dipoles are uniformly distributed throughout the receiving antenna, so the microwave beam intercepted in a local region of the receiving antenna is immediately converted into dc. The collection and rectification of microwaves with a receiving antenna based on this principle has the advantages that the receiving antenna is fixed and does not have to be pointed precisely at the transmitting antenna. Thus, the mechanical tolerance in the construction of the receiving antenna can be relaxed. Furthermore, the illumination distribution of the incoming microwave radiation need not be matched to the radiation pattern of the receiving antenna; therefore, a distorted incoming wavefront caused by non-uniform atmospheric conditions across the antenna will not reduce efficiency.

The amount of microwave power that is received in local regions of the receiving antenna can be matched to the power-handling capability of the solid-state diode microwave rectifiers. Any heat resulting from inefficient rectification in the diode circuits can be convected by the ambient air in the local region of the receiving antenna with atmospheric heating similar to that over urban areas. Only about 10% of the incoming microwave beam would be lost as waste heat. The low thermal pollution achievable by this process of rectification cannot be equaled in any known thermodynamic conversion process.

The rectifying elements in the receiving antenna can be exposed to local weather conditions. The antenna can be designed so that sunlight would still reach the land beneath it, with only a fraction lost due to shadowing. Thus the land could be put to productive use.

16

.f. Power Output Levels

Bounds can be placed on the range of potential SSPS power levels as shown in Figure 6. The SSPS design can be adjusted to provide from 4 to 40 mW/cm² of rectified power at the receiving antenna. This range of power level postulates a receiving antenna diameter of 10 to 20 km and a transmitting antenna diameter of 1 to 2 km. An idealized Gaussian distribution was chosen to establish the transmitting and receiving antenna diameters. There is an additional cutoff established by the inability of the transmitting antenna's passive thermal control system to reject the waste heat of the microwave generator when the microwave power density of the transmitting antenna rises above 4.13 W/cm². Thus, in principle, an SSPS could be designed to generate electrical power on Earth at power outputs ranging from about 2,000 to 20,000 MW.

It is likely that a narrower range of power output, ranging from 3,000 to 15,000 MW will be more effective. A nominal power output level of 5000 MW at the receiving antenna falls about in the middle of the range of interest represented by the shadowed area in Figure 6, and therefore it was chosen as representative for the SSPS baseline design. The overall capacity of the future transmission grid system will place an upper boundry on the SSPS power output to allow for the possibility that one SSPS has to be taken out of service.

SSPS Flight Control

Although the SSPS is orders-of-magnitude larger than any spacecraft yet designed, its overall design is based on present principles of technology. Thus, its construction and the attainment of a 30-year operating life require not new technology, but substantial advances in the state of the art.

The SSPS structure is composed of high-current-carrying structural elements whose electromagnetic interactions will induce loads or forces into the structure. Current stabilization and control techniques are capable of meeting the requirements of spacecraft now under development. Most of these spacecraft have comparatively rigid structures and the amenable to control as a single entity by reaction jets or momentum storage devices. But the large size of an SSPS suggests that new structural and control system design approaches may be needed to satisfy orientation requirements. This study, however, indicates that present analytical techniques/tools are adequate and that an SSPS can be controlled to better than ± 1 degree. Low-thrust, ion propulsion systems appear promising for SSPS control, because their performance characteristics are compatible with the potential lifetime required of the SSPS.

Earth-to-Orbit Transportation

A high-volume, two-stage transportation system will be required for an SSPS: (1) a low-cost stage capable of carrying high-volume payloads to low-Earth orbit (LEO); and (2) a high-performance stage capable of delivering partially assembled elements to synchronous or some intermediate orbit altitude for final assembly and deployment. The factors affecting flight mode selection include payload element size, payload assembly techniques, desirable orbit locations for assembly, time constraints, and requirements for man's participation in the assembly. The choice of transportation system elements includes currently planned propulsion stages and advanced concepts optimized for an operational SSPS system. Minimum-cost transportation combinations will have to be identified which can fulfill the requirements for SSPS delivery, assembly, and maintenance for an operational system. The challenge of an SSPS capable of generating 5,000 MW of power on Earth is to place into orbit a payload of about 25 million pounds and propellant supplies for station-keeping purposes of about 30,000 pounds per year.

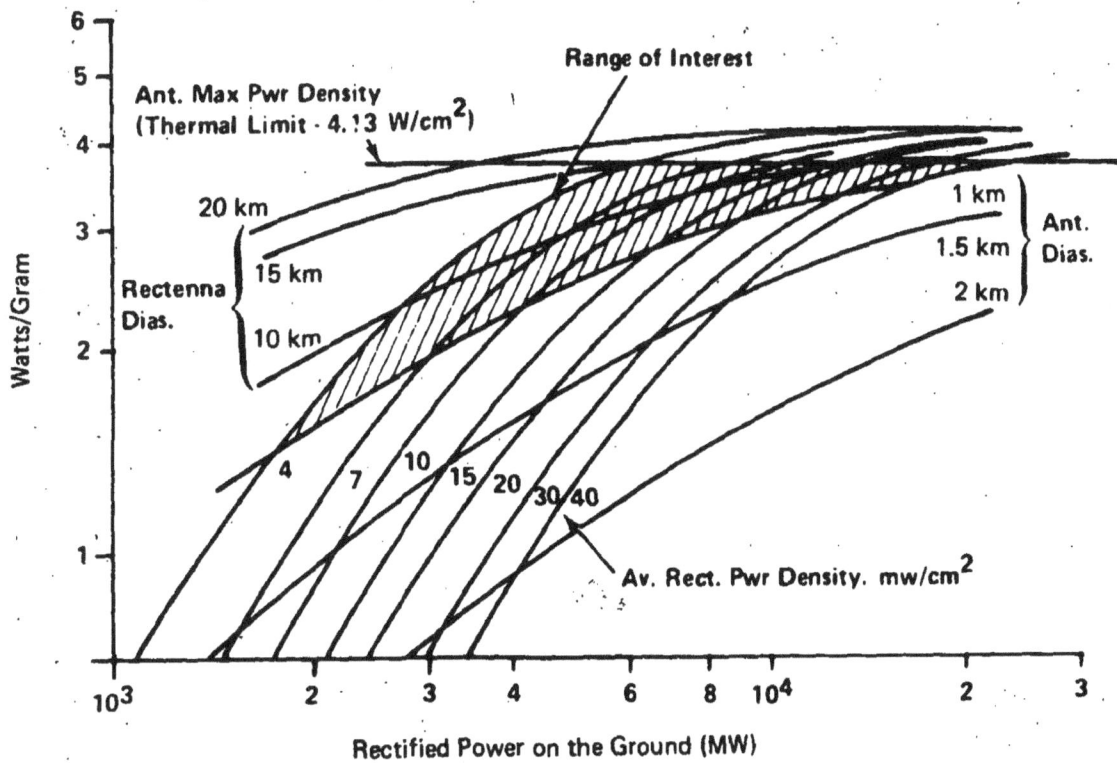

FIGURE 6. — POTENTIAL SSPS POWER LEVELS

STRUCTURE AND CONTROL TECHNIQUES

Structural and Dynamic Analysis of the SSPS

Summary. – The purpose of this task was to perform structural and dynamic analyses of the SSPS structure for the purposes of:

- Providing elastic characteristics (natural frequencies, generalized masses and mode shapes) of the structure for use in an analytical investigation of the elastic coupling between the SSPS attitude control system and the spacecraft's structural modes;

- Determining deflections and internal member loads resulting from the various flight loading conditions to verify structural integrity;

- Assessing desirable material characteristics; and

- Identifying desirable technology developments.

Results of this task will be applied to the design of structure and attitude control systems for very large-area, light-weight space structures represented by the SSPS. The SSPS will require the design of a structure that can not only support the solar cell blankets, concentrator mirrors, and transmission bus/structure for the various flight loadings, but also one that can be both assembled and controlled in space.

To investigate the structural and dynamic aspects connected with controlling such a structure in space required:

- The establishment of a baseline configuration, a structural math model, a dynamic math model, and flight loading conditions;

- The input of flight and control system loadings into the structural model to determine internal loads and structural deflections;

- Assessment of the baseline configuration for internal loads and deflections, as well as candidate material characteristics; and

- Identification of areas requiring future analysis and the analytical tools needed for analysis.

Baseline Configuration.– The SSPS characteristics and the baseline configuration used in this task are shown in Figures 7, 8, and 9. The main structural framework for each solar array consists of a large-diameter coaxial mast transmission bus, four transverse DC power buses, and non-conductive struts. Shear loads are transmitted by tension-only wires. Structural continuity between the two solar arrays is supplied by the mast and by non-conductive structure running

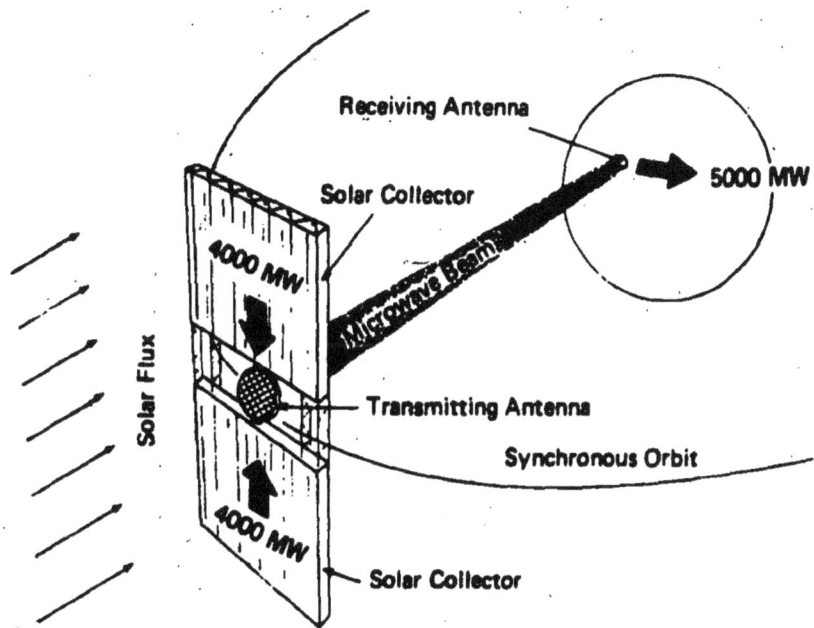

Receiving Antenna
Solar Collector
5000 MW
4000 MW
Solar Flux
Microwave Beam
Transmitting Antenna
Synchronous Orbit
4000 MW
Solar Collector

FIGURE 7. — SSPS CHARACTERISTICS

20

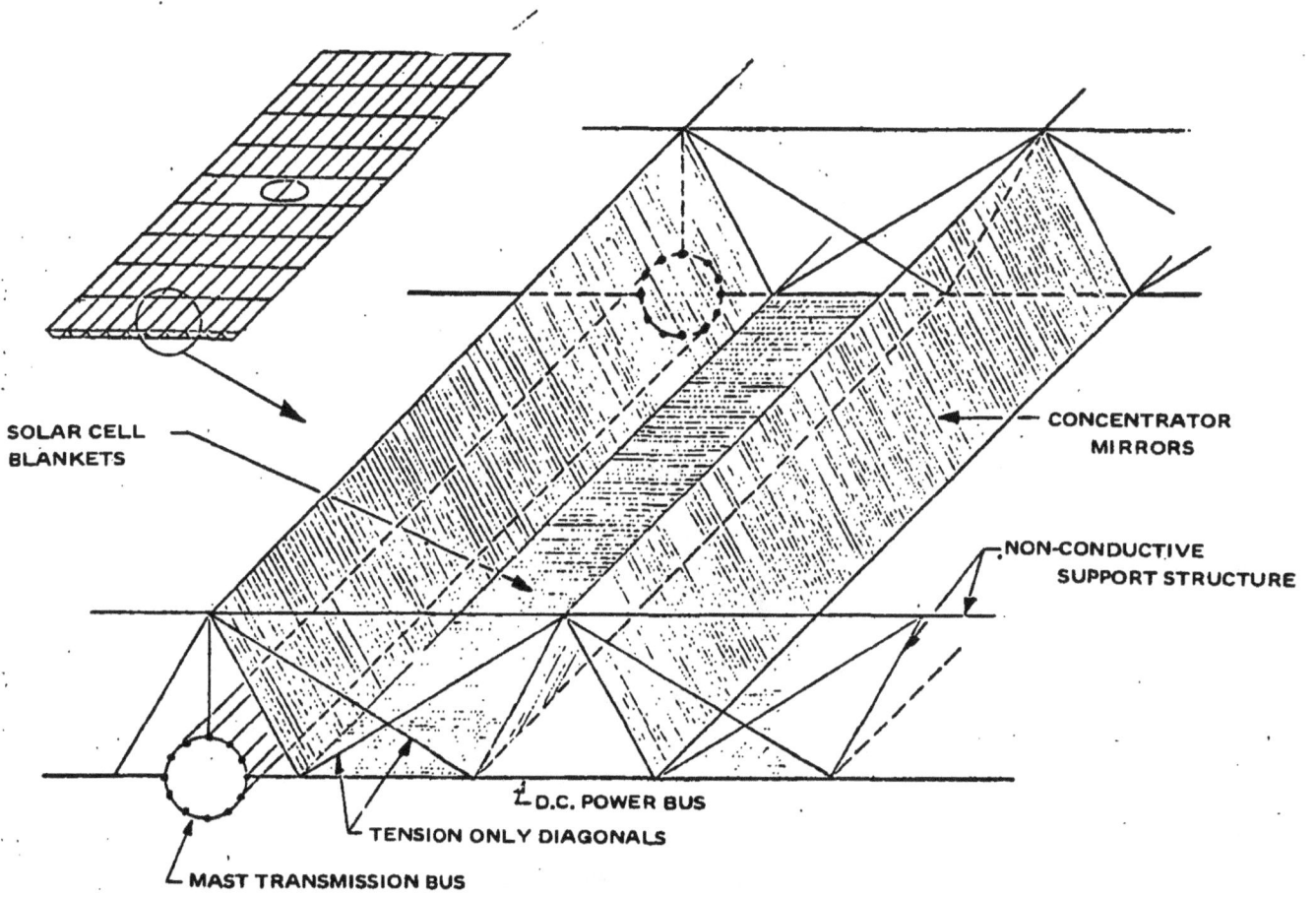

SOLAR CELL BLANKETS

CONCENTRATOR MIRRORS

NON-CONDUCTIVE SUPPORT STRUCTURE

D.C. POWER BUS

TENSION ONLY DIAGONALS

MAST TRANSMISSION BUS

FIGURE 8. — SOLAR COLLECTOR CONFIGURATION

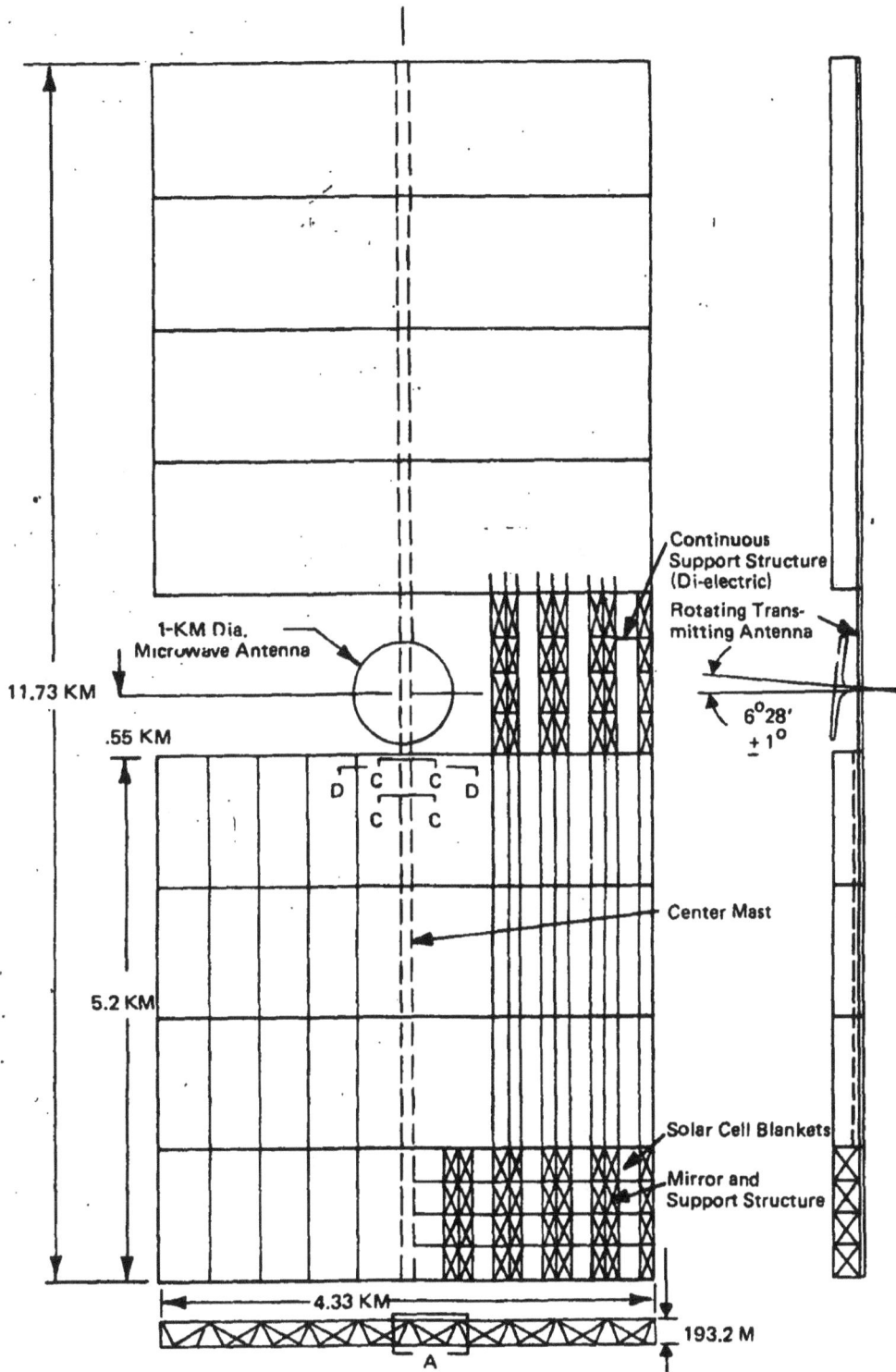

FIGURE 9. – SSPS BASELINE CONFIGURATION

Detail A – B$^+$ Buss/Beam

B$^-$ Buss/Beam

Track
(Non Conduct)

Center Beam

Rotating Mast

Section C-C
Typical Bearing

"Finger" Brushes

100M
Dia.

Section D-D
B$^-$ Power Transfer
B$^+$ Similar

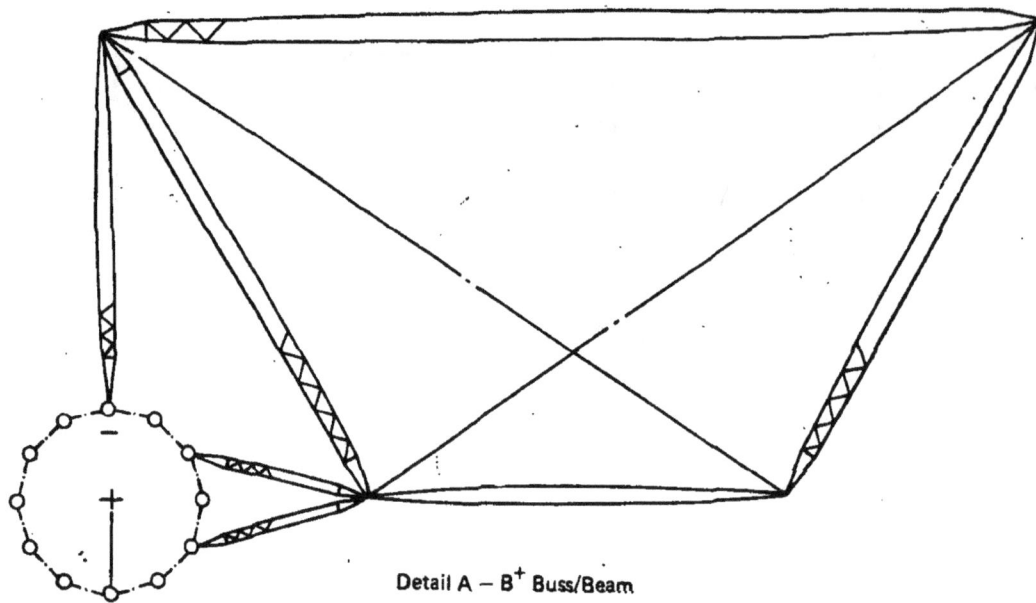

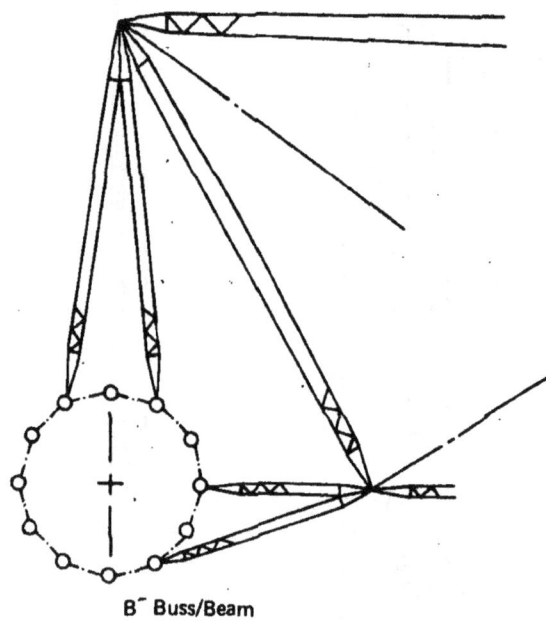

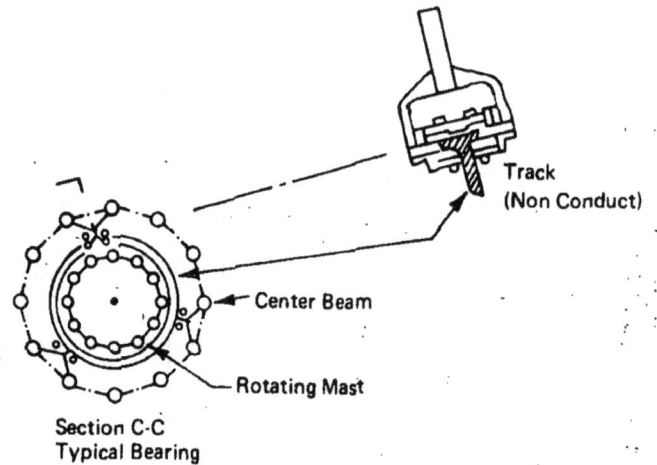

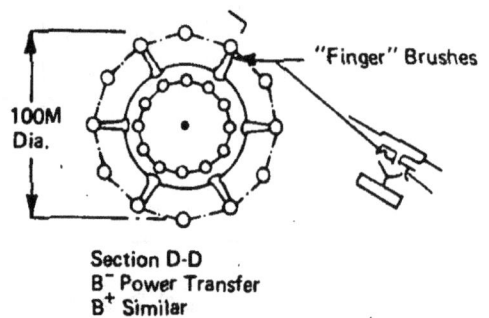

FIGURE 9. – SSPS BASELINE CONFIGURATION
(Concluded)

23

outboard of the antenna. To achieve continuous pointing of the MW antenna to the ground station (rectenna), independent rotation about the X-axis is required between the solar arrays and antenna. To accomplish this, the mast is divided into three segments – two solar array sections and an antenna section. Rotary joints at the points where the segments join permit this independent rotation. These joints transmit mast bending and axial loads, but no torsion. All twisting forces acting between the two solar arrays are transmitted by the non-conductive structure acting as torque cells. In the baseline configuration, the electrical power-conductive structure was considered capable of also carrying structural loads.

Electrical requirements for the power-conducting bus/structure call for a structure of large diameter and thin-wall tubing for heat-dissipation. Because of the resulting high diameter-to-wall thickness ratio, this type of configuration is structurally unacceptable. Thermal requirements (uniform heating of shape) also eliminate this concept. One possible attempt at solving the problem from a structural point of view is shown in Figure 10. This concept starts with a basic building element. By joining together many of these building elements bigger struts are created. This concept could be considered for all compression struts. The basic building element has to be light-weight, very easily manufactured, and have structural integrity against column failure and local instability.

In this baseline configuration, 6061 aluminum alloy was selected for the bus/structure because of its good structural properties and electrical conductivity. All the remaining structure in the plane of the four transverse DC power buses was considered to be also of 6061 aluminum to minimize the thermal warping that would result if dissimilar materials were used in this area. In the selection of the material to use for the structure connecting the solar arrays, the MW transparency of the material had to be considered, because this structure lies in the path of the microwave beam. A mica-glass ceramic was therefore used in this area for the baseline configuration. The remaining SSPS structure was the area that permitted the greatest variation in material selection. This allowed the investigation of various material properties without actually selecting any particular material. Since good structural stiffness was one of the principal requirements, an initial modulus of elasticity with a value equal to that of mica-glass ceramic was selected for the basic solar array structure. This value was used because it approximated the properties found in several common materials now used in the aerospace industry.

Structural Mathematical Model.– The structure defined in the baseline configuration was idealized into a structural mathematical model. The method used was based upon the finite-element method of structural analysis which assumes that every structure may be idealized into an assemblage of individual structural components or elements. This idealized structure is then analyzed and the results used to predict the behavior of the actual structure.

Verification of the idealization was obtained by utilizing two separate methods. The first method used the Grumman Automated Structural Analysis (ASTRAL) – Comprehensive Matrix Package (COMAP) program (defined in Reference 27). The second method used NASA's NASTRAN program (defined in Reference 28). Both programs employed identical idealization of structure. The ground rules and assumptions used in both models are listed below. To reduce the computer time required to run both problems, the structure was considered symmetrical about the

24

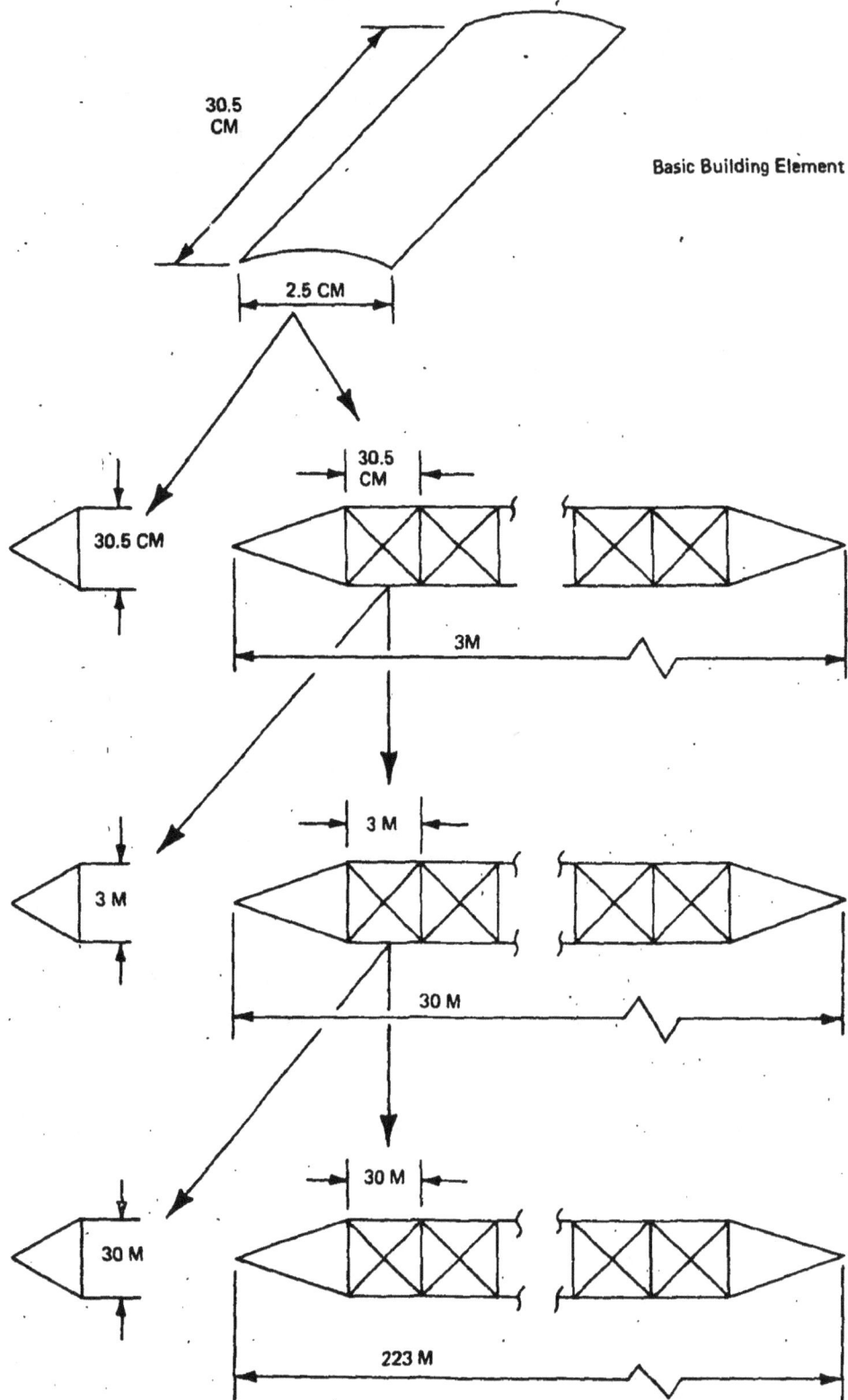

30.5
CM

2.5 CM

Basic Building Element

30.5
CM

30.5 CM

3M

3 M

3 M

30 M

30 M

30 M

223 M

FIGURE 10. — CONSTRUCTION OF COMPRESSION STRUT

antenna centerline perpendicular to the mast and therefore only half the structure had to be idealized. Symmetrical and anti-symmetrical models were then run using both methods of solution. Ground rules and assumptions used in both models are presented below:

- Structure symmetrical about antenna centerline perpendicular to mast.

- Analysis uses only half structure.

- Antenna is included as rigid body and lies at centerline of mast.

- Antenna has 5 degrees of freedom, no rotation about X-axis.

- Mast idealized as consisting of multiple beams having bending stiffness, but no torsional capability.

- Mast moments of inertias based on six current elements per polarity.

- All other support structure idealized as axial loaded struts.

- Areas of mast and solar array buses based on power distribution system operating temperature of 38°C (100°F).

- Solar array elements lie in plane of blankets.

- Total cross-sectional area of non-conductive struts is 0.27 in.²

- Tension-only wires replaced by single tension/compression struts; cross section area is 0.01 in.²

- Model representing half structure consists of 684 members and 232 nodes.

- Satellite weight distributed as lumped masses at node points.

Table 3 gives a summation of the weights and inertias used in the structural analysis.

Dynamic Mathematical Model — A dynamic analysis was performed on the symmetrical and anti-symmetrical structural models using both ASTRAL/COMAP and NASTRAN. Both methods are normal-mode solutions using large digital computer programs. Solution consists of the natural frequencies of the system, relationship between the degree of freedom displacements or mode shapes, and generalized model masses and stiffnesses. The solution using the NASTRAN method retained virtually all the degrees of freedom associated with the 232 grid points, with each grid point having 3 degrees of freedom. The ASTRAL/COMAP solution had to be reduced to about 250 degrees of freedom because of the limitation of the size of the problem that the program can handle. For checking purposes, solutions using both ASTRAL/COMAP and NASTRAN were

TABLE 3

SSPS STRUCTURAL MODEL WEIGHTS

	Weight (kg x 10^6)	Weight (lb x 10^6)
Solar Array	8.12	17.89
Blankets	6.11	13.47
Concentrators	1.02	2.25
Bus/Structure	0.41	0.90
Mast	0.58	1.27
MW ANTENNA	1.98	4.37
ROTARY JOINTS	0.32	0.70
Total Weight	10.42	22.96

	Inertias (kg-km^2)	Inertias (slugs-ft^2)
MW ANTENNA		
I_{YY}	134,000	9.87×10^{10}
I_{ZZ}	247,500	18.17×10^{10}

initially derived for the first four modes in both the symmetrical and anti-symmetrical models. Comparison of the results showed fairly reasonable agreement between the two methods. Differences in results are attributable to loss of accuracy in the ASTRAL/COMAP approach, due to the relative magnitude of the numbers used in the geometry and member areas. NASTRAN has the capability to handle numbers of much larger magnitude than ASTRAL/COMAP. Also, the reduction in the problem degrees of freedom to accommodate ASTRAL/COMAP adds to its slight inaccuracy. Because of NASTRAN's greater accuracy and faster problem setup time when changing basic input data, this method of analysis was used for the remainder of the study effort.

After verifying that initial results obtained from the dynamic model appeared accurate, additional dynamic solutions were obtained. The first 15 symmetrical modes and the first 14 anti-symmetrical modes were derived and the results tabulated in Tables 4 and 5.

Input of Elastic Body Characteristics into Attitude Control System Analysis.— Elastic body characteristics derived from the dynamic math model were inputted into the attitude control system analysis in References 29 and 30. The purpose of those studies was to perform the following tasks:

TABLE 4

DYNAMIC MODEL — ELASTIC BODY CHARACTERISTICS*

SYMMETRICAL

Mode	Frequency (Hz)	Generalized Mass (weight/12 G) (lb-sec^2/in.)	Generalized Stiffness (lb/in)
1	4.090	2726	.139
2	4.360	7898	.457
3	6.267	8661	1.036
4	9.457	3346	.912
5	12.056	5971	2.644
6	12.611	3891	1.884
7	13.972	5797	3.447
8	14.317	4881	3.049
9	16.362	3400	2.773
10	18.119	1346	1.346
11	18.320	6270	6.411
12	19.793	3542	4.227
13	26.719	3387	7.365
14	28.220	5035	12.214
15	29.380	4877	12.824

*All values are for half structure

TABLE 5

DYNAMIC MODEL – ELASTIC BODY CHARACTERISTICS*

ANTI-SYMMETRICAL

Mode	Frequency (Hz)	Generalized Mass (weight/12 G) ($lb-sec^2/in.$)	Generalized Stiffness (lb/in.)
1	.187	2,750	.00292
2	6.757	3,265	.454
3	8.431	14,396	3.118
4	10.199	6,662	2.111
5	10.933	3,879	1.412
6	13.525	4,263	2.375
7	14.454	6,051	3.850
8	14.486	5,834	3.729
9	16.542	3,901	3.251
10	18.410	5,772	5.961
11	23.911	963	1.678
12	24.790	2,758	5.163
13	25.852	11,980	24.386
14	27.353	1,307	2.978

*All values are for half structure

- Define an attitude control system for the SSPS and investigate analytically the elastic coupling between the control system and the spacecraft's structural modes.

- Evaluate the performance of the baseline SSPS and perform a parametric sizing study to determine the influence of structural flexibility upon system performance.

Results of those studies are presented in References 29 and 30.

Establishment of Flight Loading Conditions.— The SSPS nominal orientation in synchronous orbit is defined in Reference 31. External forces acting on the satellite will cause deviations in the nominal orbit which, unless corrected by the spacecraft attitude control system, will cause pointing errors in the solar collectors and antenna. Pointing errors of the antenna can be corrected by a combined mechanical and electronic system and do not require any corrections of the SSPS main structure. For the purposes of this study, allowable deviation angles in the solar collector pointing accuracy were limited to ±1.0 degree about all axes. Mass expulsion actuators are used to control angular deviations and orbital drift.

The external forces acting on the SSPS have four sources: aerodynamic, magnetic, solar pressure, and gravity gradient (Figure 11). The external forces associated with these conditions are given in Reference 31. At synchronous altitude, the aerodynamic forces are negligible. The magnetic forces acting on the SSPS due to the interaction of the magnetic fields of the Earth and the SSPS are shown to be small. The total solar pressure force acting on the satellite is 224 newtons (50 lbs) with small resultant torques. The largest external torques acting on the satellite are caused by gravity-gradient effects. Values of these torques are a direct function of the angular offset from the nominal orientation. Values for the gravity-gradient torques are given in Reference 31, and the control actuator forces required to correct for these torques are given in Reference 30.

The passage of electrical currents along the bus/structure elements generates forces in these members. The magnitude of these electromagnetic forces is given in Reference 32. These forces are internal and are self-balancing.

A study was conducted to determine the force created by the radiation of electromagnetic energy from the SSPS antenna (in effect this is a "recoil" from the MW beam). The study showed that the total force normal to the antenna was 1.8 kg (4 lbs).

As the Earth rotates about the Sun in the ecliptic plane, the Earth casts a circular shadow of approximately its own diameter. Every satellite in equatorial synchronous orbit crosses this shadow daily during a 45-day period, twice per year at the time of the vernal and autumnal equinoxes. The time duration spent by the SSPS in this eclipse varies from near zero to a maximum of about 72 minutes at the equinox date. As the satellite enters the eclipse, it experiences a temperature drop of about 270°K. The thermal transient is very rapid at the beginning of this period, and a satisfactory method of assessing the loading and dynamic effects of rapid dimensional fluctuations, induced by an eclipse interval, remains to be analyzed. The overall structural response of the SSPS during this period and after reentry into the sunlight represents a significant future study area involving the dynamic effect on large structures of rapid thermal transients.

30

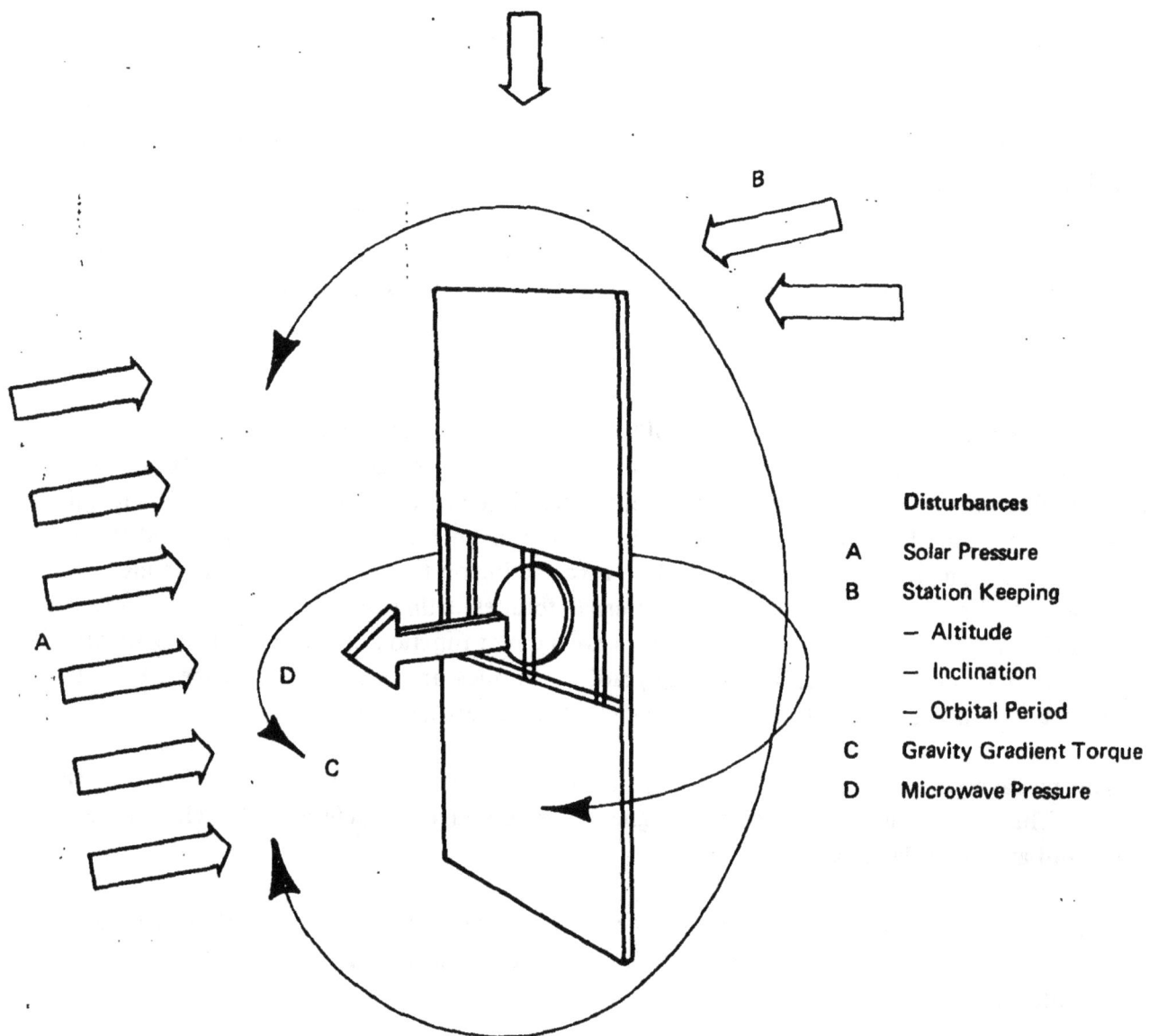

Disturbances

A   Solar Pressure

B   Station Keeping

  — Altitude

  — Inclination

  — Orbital Period

C   Gravity Gradient Torque

D   Microwave Pressure

FIGURE 11. — SSPS DISTURBANCE FORCES AND TORQUES

When the SSPS is eclipsed by the Earth, the temperature of the blankets and mirrors will drop more rapidly than the supporting structure. The result is that a cold blanket or mirror will coexist with a warm structure. As the SSPS comes out of eclipse, the opposite condition exists. These differences in temperature cause differences in thermal distortions between the blankets and mirrors and their supporting structure. To compensate for the relative differences in distortions and also to keep the blankets and mirrors from wrinkling, the use of pretensioning devices between the blankets and mirrors and their supporting structure is anticipated. The magnitude of load that these devices supply and their locations are given in Reference 33.

Input of Flight and Control System Loadings into Structural Model to Determine Internal Loads and Structural Deflections.– Unit forces were applied to the structural math model at the grid points that correspond to the various attitude control actuator locations. These control forces were balanced by inertia forces acting at the various mass points. Internal member loads and structural deflections were derived for these unit forces using the NASTRAN Static Analysis Solution given in Reference 28. Solutions for 100-kg control forces applied at actuators located at the outboard end of the mast are given in Figures 12 and 13.

Assessment of Baseline Configuration for Internal Loads and Deflections. – The primary purpose of this portion of the study was to investigate whether a large-area, light-weight space structure (represented by the baseline SSPS configuration) could be controlled in space, and then to check the structural integrity of the spacecraft for the resulting actuator control forces. The results of the performance evaluation of the baseline system performed in Reference 30 has shown that the SSPS can be controlled to well within a ±1 deg pointing accuracy with end thrusters of only 4.5 kg (10 lbs). These thruster loads counterbalance the gravity-gradient disturbance torques discussed earlier. Internal member loads and structural deflections resulting from these forces are negligible.

However, in addition to attitude control forces, additional external control forces are required for corrections due to orbital drift resulting from such disturbances as solar pressure. The magnitude of these control forces is dependent upon the system's duty cycle and corresponding propellant limitations. No attempt was made to define the recommended duty cycle for this mission. Instead, a structural analysis was performed [Reference 34] which indicated that, to prevent localized bending of the solar array structure from exceeding ±1 deg, the end thruster force must not exceed 303 kg (667 lbs). Since the orbital correction force, dependent upon its duty cycle, could exceed this value, the results of the structural analysis show that the system's duty cycle must be selected so that the control force does not exceed 303 kg.

The end thruster force of 303 kg in the Z-direction is therefore the critical loading condition acting on the SSPS structure. The maximum internal load resulting from these control forces was calculated in Reference 34 to be 290 kg (640 lbs) limit. This load occurs in the carry-through structure surrounding the MW antenna. Using a structural model area of 1.74 cm^2 (0.27 in.^2) gives a maximum compressive stress of 167 kg/cm^2 (2375 psi). Ignoring the local buckling allowables, to obtain a column allowable of 167 kg/cm^2 for a member having a modulus of elasticity (E) of $0.64 \times 10^6 \text{ kg/cm}^2$ (9×10^6 psi), would require that the L/ρ of this column not exceed 194. This L/ρ value implies that a column to support this load could probably be constructed.

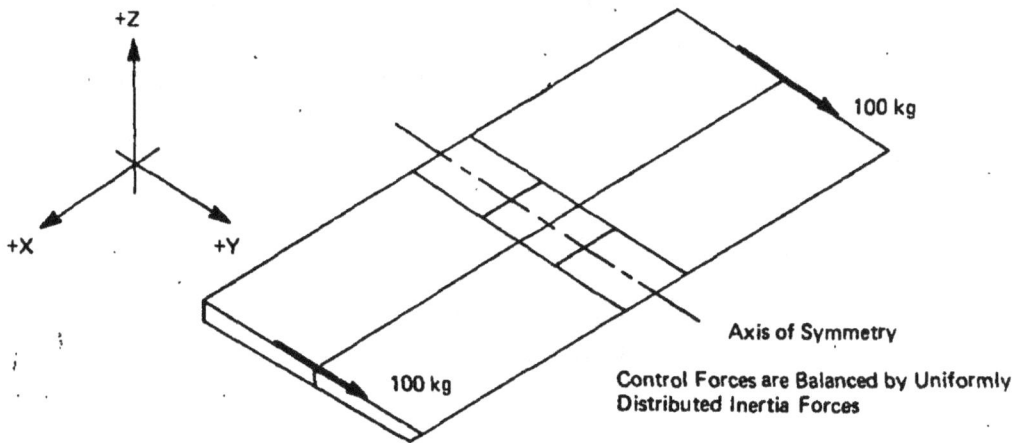

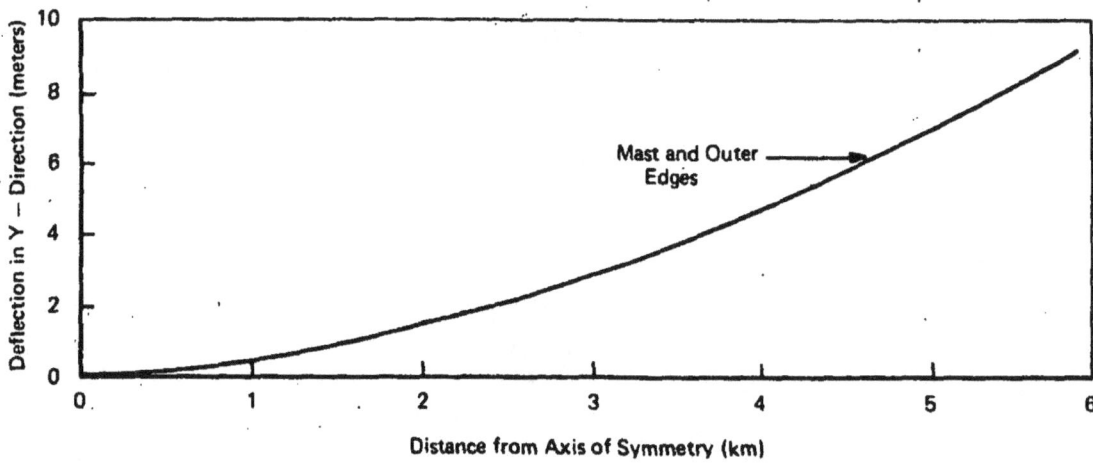

Maximum Member Loads

Mast (Bending) = 9.6×10^4 kg — meters (8.36×10^6 in.-lb)

D.C. Power Bus Members (Axial) = 33 kg (72.8 lb)

Diagonals (Axial) = 20 kg (44 lb)

Non-Conductive Members (Axial) = 26.4 kg (58.1 lb)

FIGURE 12. — SSPS DEFLECTIONS AND MEMBER LOADS FOR UNIT CONTROL FORCES ACTING IN THE Y DIRECTION

Control forces are balanced by
uniformly distributed inertia forces:

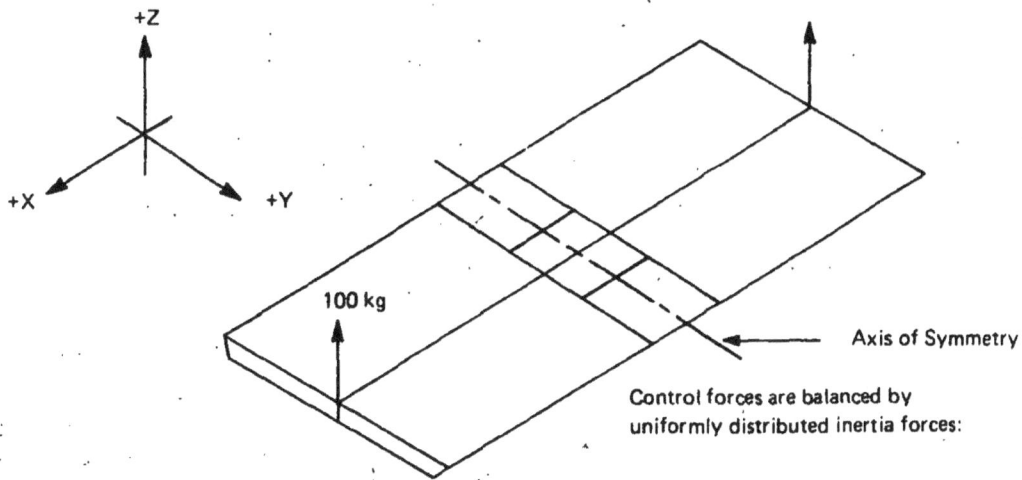

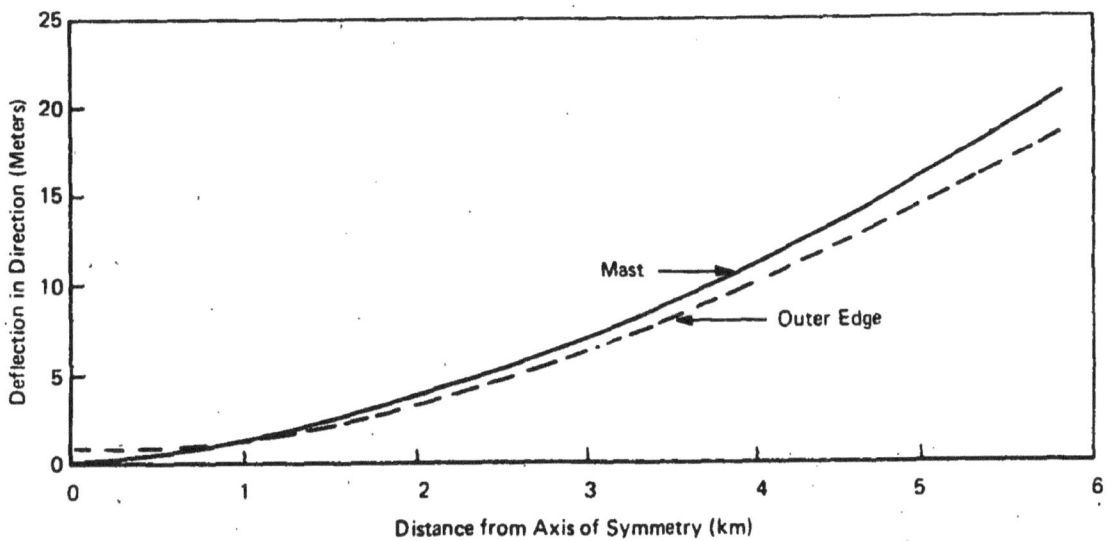

Maximum Member Loads

Non-Conducting Members (Axial) = 96 kg (211 lb)

Mast (Bending) = 20.8×10^4 kg – Meters (18.1×10^6 in-lb)

D.C. Power Bus Members (Axial) = 146 kg (322 lb)

Diagonals (Axial) = 16.5 kg (36.5 lb)

FIGURE 13. – SSPS DEFLECTIONS AND MEMBER LOADS FOR UNIT
CONTROL FORCES ACTING IN THE Z DIRECTION

34

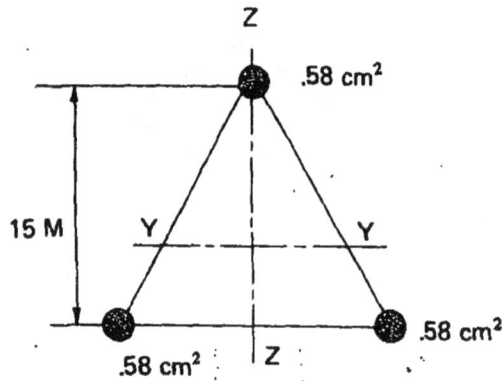

Total Area = 1.74 cm²

$I_{ZZ} = 652,000$ cm⁴

$\rho_{ZZ} = 612$ cm

$L = 665$ m

$L/\rho < 194$

An analysis of the configuration using the pretensioning loads on the blankets and mirrors shows the critical loaded area occurs where two mirrors are connected to a strut running parellel to the X-axis. The resultant vertical load on the structure is 2(0.866) (0.91) = 1.59 kg (3.5 lb) every 10 meters. Applying these forces to a beam 325 meters in length with a $I_{YY} = 760,000$ cm⁴ gives a bending stress of 239 kg/cm² (3400 psi). This moment can occur concurrently with an axial load in the strut of 268 kg (590 lb) compression. The axial load results from applying the 303-kg thruster force. The combination of axial load and bending moment requires that the strut now be analyzed as a beam column. Since the axial load is less than that used in the earlier column analysis, the strut — acting as a column only — is adequate. The allowable bending stress of the strut is dependent on the local buckling allowable of the strut which, in turn, requires knowing the configuration of the section. Since the strut configuration is not known at this time, a complete analysis of this beam column cannot be included. What can be demonstrated, however, is that the applied forces are not excessive for the baseline configuration and adequate light-weight structure can be designed.

The forces generated in the bus/structure by the passage of electrical currents along the integral conductors are presented in Reference 32. Electromagnetic forces between parallel bus elements are very low because of their wide separation; the maximum forces occurring in the mast can be very conservatively estimated at 3 gm/cm length. This force produces bending in the individual mast elements. A complete analysis of this section cannot be performed until a detailed configuration is finalized. The applied loading is very small and bending lengths can be minimized by introducing insulated internal bracing into the mast to support the electromagnetic forces which will then be self-balancing.

Assessment of Candidate Material Characteristics.— Material criteria for the SSPS configuration include the following:

- Modulus of elasticity

- Coefficient of thermal expansion

- Weight

- Electrical conductivity

- Space longevity – Outgassing – Fatigue allowables – Radiation resistance

- Producibility

- MW transparency

In the discussion of the baseline configuration, it was shown that in certain material selections (such as 6061 aluminum for the bus/structure and mica-glass ceramic for the carry-through structure surrounding the MW antenna), because of very particular requirements in these areas, very little variation can be permitted in the materials selected. Other areas in the configuration, such as the non-conductive struts, did permit variation in material selection.

In analyzing a structure for its elastic body characteristics, the only criterion from the above list to be considered is the material's modulus of elasticity. For the non-conductive struts, variations in their modulus of elasticity were inputted into the symmetrical dynamic model and the effects on the total structure's natural frequencies were plotted in Figures 14 and 15. As can be seen in the curves, when considering symmetrical modes, variations in the non-conductive struts' modulus of elasticity affect only torsional modes; even with these modes, increasing the value of the modulus over that used in the baseline configuration has practically no effect on the total structure frequencies. In further analyses, variations of the modulus should be inputted into the anti-symmetrical dynamic model. It can be estimated, however, that variations would probably again not have much effect on the bending modes. However, since torsion is more predominant with anti-symmetrical modes than with symmetrical, a greater change in frequencies can be anticipated than is shown in Figure 14.

Identification of Areas Requiring Further Structure and Dynamic Analysis.– The following items are areas in which future analytical analysis is needed:

a. Improved Method for Reduction of Number of Variables Used in
   Equations of Motion

It is likely that the attitude-control jets will excite a number of high-frequency vibration modes as well as the lower-frequency and rigid-body modes. These motions, in combination, may result in waves which emanate from each jet and damp out as they proceed through the structure. To predict the dynamic behavior of the structure accurately, it appears that an unusually large number of modes will be required; thus, the computer time and storage requirements would be excessive. For these reasons a study is recommended to determine more effective dynamic-analysis methods for large flexible space structures. The improved techniques developed in this study provide increased confidence in the capability of the control system to achieve the required control.

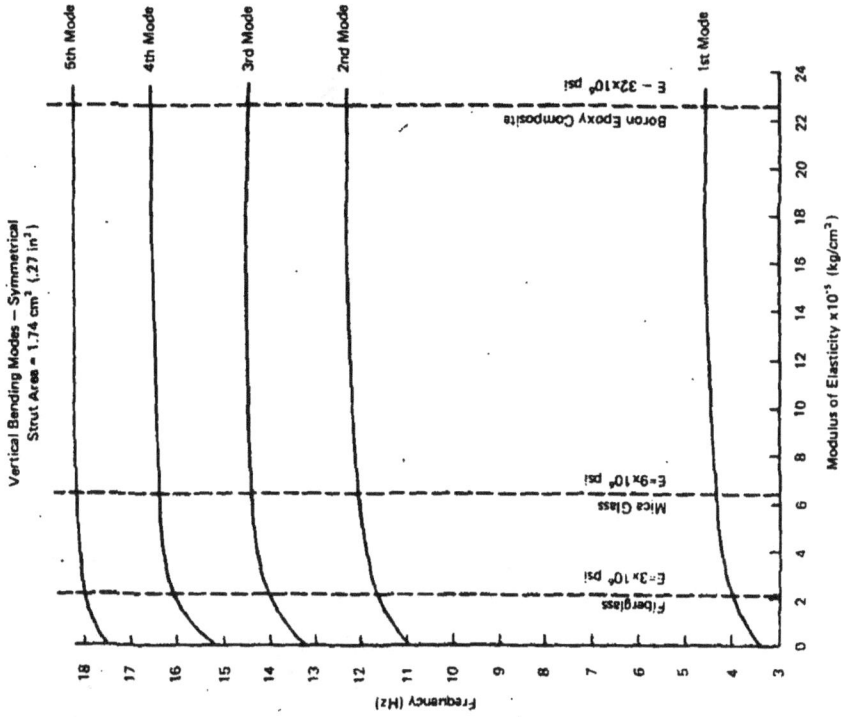

FIGURE 15. – SSPS NATURAL FREQUENCIES VERSUS MODULUS OF ELASTICITY OF NON-CONDUCTIVE STRUTS

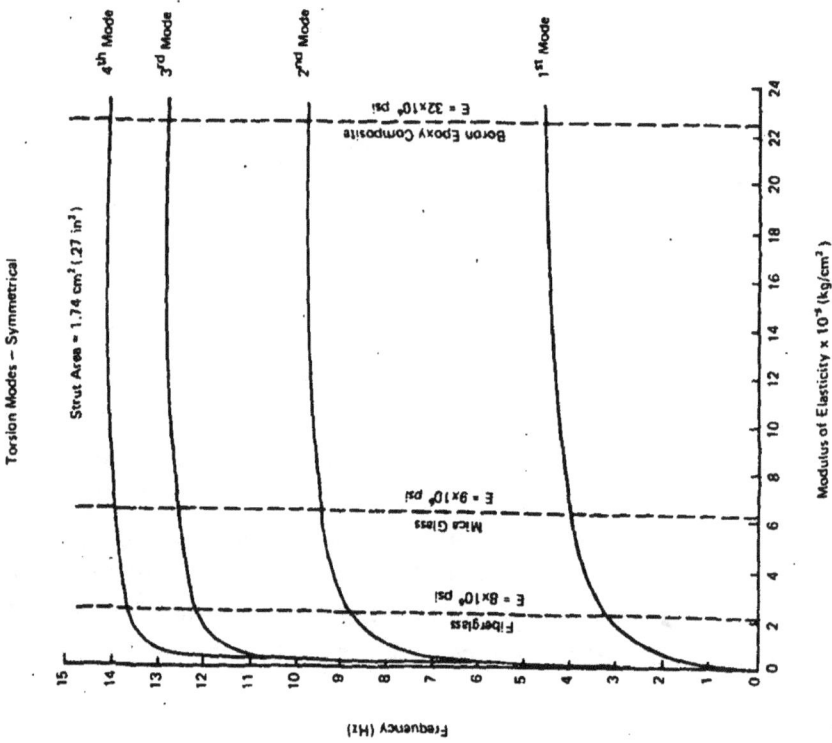

FIGURE 14. – SSPS NATURAL FREQUENCIES VERSUS MODULUS OF ELASTICITY OF NON-CONDUCTIVE STRUTS

37

In the recommended study, generalized coordinates such as components of wave functions would be considered as a substitute for higher order modes. Various sets of functions would be investigated, and the accuracy of the solutions obtained would be compared by substitution into the equations of motion. The study would proceed as follows:

- Treat simple structures, such as beams and plates,

- Assess accuracy and compare with modal approach,

- Based on the above results, develop methods to generate the desired functions for complex lumped mass structures such as the power station, and

- Evaluate accuracy and compare with modal approach.

The end-product would be a computer program which would automatically generate the required expansion functions.

b. Time-History Program with Accurate Evaluation of Gravity-Gradient Excitation

To study, in detail, the control of a flexible structure in a gravity-gradient field, it is necessary to determine accurately the difference between the gravity force and the orbital centrifugal force at each mass point. In many existing computer programs these effects are computed and then subtracted; however, the effects are nearly equal, and it is the small difference which is of consequence. This procedure is, therefore, considered too inaccurate to be of value. To improve the procedure significantly, the gravity and orbital centrifugal effects should be expanded in a series, analytically subtracted, and the result should be programmed in a general time-history structural program.

This computer program should be sufficiently general so that structures of any shape and mass distribution may be treated. In addition, to include gravity effects properly, it should have the capability of including a large variety of control systems as well as a general set of externally applied loads. It should not rely on orthogonal functions such as modes for the reduction of coordinates, so that the above described expansion functions may be used. The resulting program would be a powerful tool for predicting the dynamic behavior of a flexible structure in space.

c. Dynamic Response to Thermal Excitation

When the SSPS enters and leaves the Earth's shadow, it undergoes severe temperature changes. The thermally induced deformations during this alternate cooling and heating may result in significant vibration of the SSPS, and an investigation of this effect is recommended.

A three-phase analysis is proposed as a practical approximate approach to accomplish this investigation. In the first phase, the temperature time-history of the satellite would be predicted using existing computer programs such as NASTRAN or Grumman's Integrated Thermal Analysis Procedure. Next, the external loads which produce deformations that are equivalent to the thermal deformations would be generated. NASTRAN or Grumman's ASTRAL-COMAP computer program system could be used to accomplish this phase. In the third place, these loads could be introduced into the new time-history program recommended in Item b. in order to determine the dynamic behavior of the satellite.

Flight Control Performance Evaluation of the Baseline SSPS

Summary.– The purpose of this task was to evaluate the flight control performance of the baseline SSPS and to perform parametric studies to determine the influence of structural flexibility upon control system performance.

The parameters of interest in this study included: structural stiffness (frequency), steady-state attitude error, control gains, control thrust levels, response time, damping ratio and control frequencies.

As a result of the flight control performance evaluation, it was determined that the pointing accuracy of the SSPS is well within the ±1 deg limit specified by the baseline requirements for the pitch, roll and yaw axes. In addition, the system's response time and percent overshoot were found to be acceptable for all three axial modes.

The results of the parametric studies showed that as the structural frequency (stiffness) decreased the system's attitude errors and response times increased. However, it was found that for as much as a 50% decrease in structural weight (25% decrease in frequency) the system's pointing accuracy was still well within specification.

The results of this study phase indicated that structurally the baseline design is sufficiently stiff to allow excellent attitude control. In fact, it could be concluded that the present baseline structure is overdesigned. However, before steps are taken to further update the baseline SSPS design it is suggested that: (1) Structural and control implications of assembling an SSPS in orbit be evaluated; (2) Analyses of the effects of rapid thermal transients induced by eclipse periods be conducted; (3) Refinements of structure/control interaction analyses be implemented to better understand the dynamic behavior of very large structures; and (4) The above efforts be focussed toward identifying the minimum weight vehicle system having acceptable structural stiffness and pointing capabilities.

Ground Rules and Assumptions. —

- Planar (Single Axis) Analysis — Spacecraft's three rotational modes are dynamically uncoupled

- Linear Analysis – Linear control system and structural model

- The SSPS can be modeled as a free-free beam

- The SSPS is structurally symmetric.

Discussion.– The objective of this study phase was to (a) evaluate the flight control performance of the baseline system, and (b) to conduct a parametric sizing study to determine the sensitivity of the system's performance to variations in structural stiffness. These tasks were performed by using the parametric performance indices developed in Reference 35 and the resluts of the structural analysis presented in Reference 36. Additional studies were also conducted using a non-real time 360-75 digital simulation of the controlled dynamics of the SSPS.

Since the overall study has assumed that the rotational dynamics of the SSPS are uncoupled, individual analyses have been performed for the pitch, roll, and yaw modes. The results of each of these evaluations are presented below.

a. Pitch Mode

Pitch mode dynamics have been defined to be rotation about the spacecraft's Y axis in the X-Z plane as shown in Figure 16.

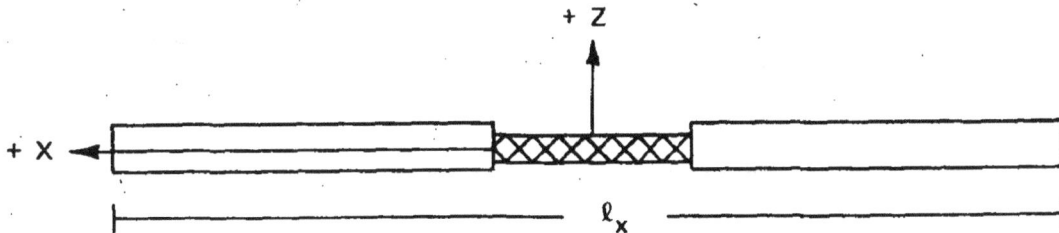

FIGURE 16.   PITCH AXIS

The baseline rigid body mass properties of the SSPS about this axis are listed in Table 6.

TABLE 6

PITCH AXIS MASS PROPERTIES

I_{YY}	Length
$123. \times 10^6$ kg-km^2 90.53×10^{12} slug-ft^2	12 km 7.46 miles

TABLE 7
PITCH AXIS BENDING MODE DATA

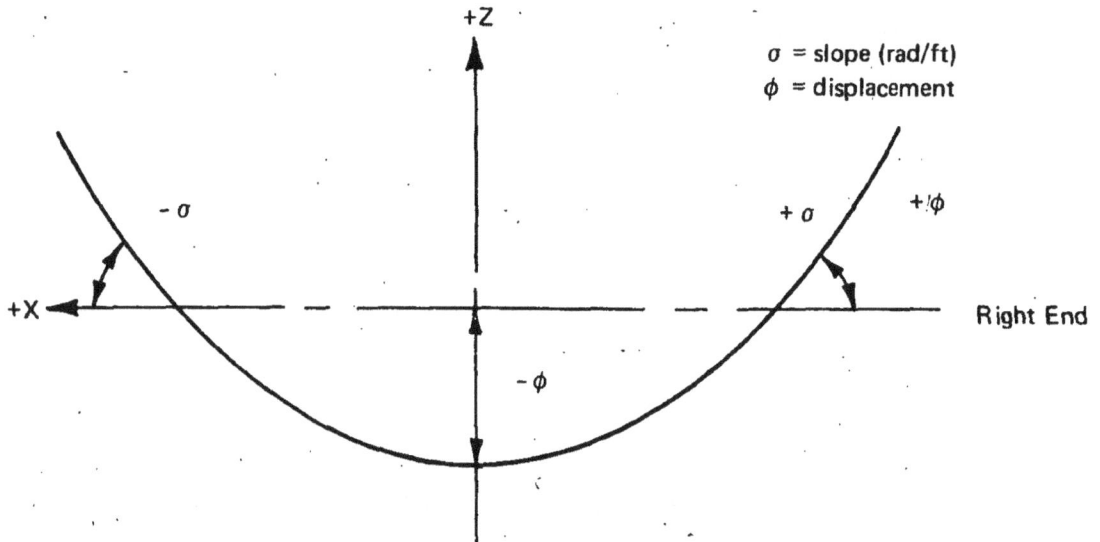

Mode	1st SYMM	1st ANTI-SYMM	2nd SYMM	2nd ANTI-SYMM
Frequency Rad/Sec.	0.0076	0.0178	0.021	0.025
Generalized Mass–Slugs	189576.	159888.	143304.	140016.
Structural Damping	0.0	0.0	0.0	0.0
Right End Normalized Displacement ϕ	+ 1.0	+ 1.0	- 0.793	+ 0.760
Right End Normalized Slope σ (Rad/Ft)	$+0.864 \times 10^{-3}$	$+0.18 \times 10^{-3}$	$- 0.191 \times 10^{-3}$	$+0.353 \times 10^{-3}$

As a result of the structural dynamic analysis, four vertical bending modes were identified for the baseline structure. The mode shapes for each of the vertical modes are shown in Figures 17 through 20. As indicated, Figures 17 and 19 show symmetric modes while the anti-symmetric modes are shown in Figures 18 and 20. The modal characteristics for these modes are identified in Table 7. The normalized modal displacements, ϕ, and the normalized modal slopes, σ, corresponding to each mode, are only identified at the ends of the axis where the control actuators and sensors are located on the baseline system (Reference 36). These data are only given for one side of the structure since structural symmetry is assumed. A more detailed discussion of these mode shapes is given in Reference 36.

Nominally the X axis of the SSPS is perpendicular to the orbit plane. An assessment of the Y axis disturbance torques in this orientation has determined that there is a constant torque, T_{dy} due to solar pressure as well as a gravity gradient torque, K_{dy} θ_y which is proportional to the off-nominal deviation angle, θ_y. Nominally this latter term is zero.

The magnitude of these effects was evaluated in Reference 31 and is repeated in Table 8.

TABLE 8

PITCH AXIS DISTURBANCE TORQUES

T_{dy}	K_{dy}
5550 N-M	176×10^4 N-M/rad
4100 ft · lb	131.8×10^4 ft · lb/rad

As a result of the controllability study (Reference 35) it was determined that the system's transient response characteristics could be approximately measured in terms of the system's rigid body damping ratio, ζ, and undamped natural frequency, ω_n. Parametric expressions for each of these terms are given in Reference 35 and are repeated in Equations (1) and (2) below:

$$\omega_n = \left(\frac{K_A \Sigma \ell}{I} - \frac{K_d}{I} \right)^{\frac{1}{2}} \text{ rad/sec.} \qquad (1)$$

$$\zeta = \frac{K_r}{2} \left[\frac{\left(\frac{K_A \Sigma \ell}{I} \right)}{\left(\frac{K_A \Sigma \ell}{I} - \frac{K_d}{I} \right)^{\frac{1}{2}}} \right] \qquad (2)$$

42

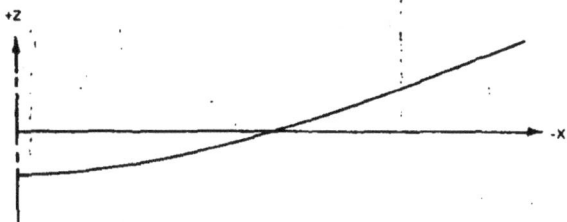

FIGURE 17. — FIRST SYMMETRIC PITCH AXIS
BENDING MODE SHAPE

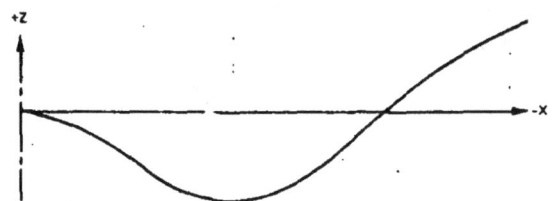

FIGURE 18. — FIRST ANTI-SYMMETRIC PITCH
AXIS BENDING MODE SHAPE

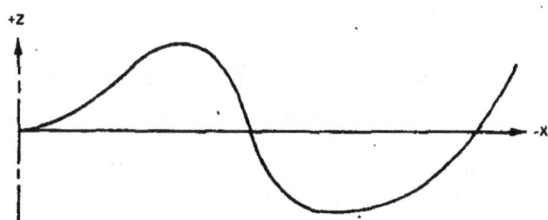

FIGURE 19. — SECOND SYMMETRIC PITCH AXIS
BENDING MODE SHAPE

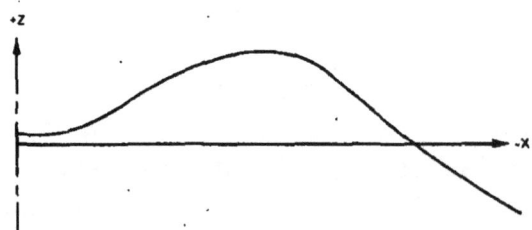

FIGURE 20. — SECOND ANTI-SYMMETRIC PITCH
AXIS BENDING MODE SHAPE

43

For the baseline spacecraft design, the system's damping ratio was selected to be 0.50, so as to limit the peak oscillations of the system's response. To separate the system's control dynamics from the structural dynamics, an undamped natural frequency for each axis was chosen to be a factor of 10 less than the corresponding lowest anti-symmetrical bending mode frequency.* For the pitch axis this results in an undamped natural frequency of 0.00178 rad/sec.

Using Equations (1) and (2) we found the baseline control parameters for the pitch axis to be:

$$K_{A_y} = 7700.0$$

$$K_{R_y} = 570.0$$

With the above set of parameter values, the dynamic performance for the pitch axis of the baseline spacecraft was evaluated using the parametric performance indices defined in Tables 1 and 5 of Reference 35.

The results of the pitch axis performance evaluation are summarized below:

$$\frac{\Delta\theta y}{\theta_{cy}} = 5.10 \times 10^{-3} \text{ rad/rad}$$

$$\frac{\Delta\theta y}{T_{dy}} = 3.5 \times 10^{-9} \text{ rad/ft - lb.}$$

$$\frac{F_{cy}}{T_{dy}} = 2.7 \times 10^{-5} \text{ lb/ft - lb.}$$

$$T_{R_y} = 3180 \text{ sec.}$$

These results indicate that for each degree the SSPS is commanded to point about the Y axis, the attitude error would be 0.005 deg. However, since the nominal orientation requires a zero off-set angle the first terms contribute a zero attitude error.

However, in attempting to maintain a nominal orientation the spacecraft must counteract disturbance torques which are tending to cause it to move off nominal. The second quantity listed above indicates that, in the presence of disturbance torques, the attitude error for the pitch mode is 3.5×10^{-9} rad/ft-lb. Therefore, for a constant disturbance torque about the Y axis of 4100 ft-lb, the steady-state, off-nominal attitude error is 1.50×10^{-5} rad (86×10^{-5} deg). To counter this disturbance the baseline control system must continuously exert 0.11 lb of thrust in coupled pairs at the extremities of the Y axis (as indicated by the third quantity listed above). In responding to this disturbance, the pitch axis dynamics will decay to within 95% of its steady-state value in 3180 sec.

---

*Only anti-symmetrical bending modes are dynamically coupled with the attitude control system (see Reference 35).

In comparison to the ± 1-deg attitude pointing requirement for the baseline system, the pitch-axis performance is well within specifications. In addition, it was determined in the structural analysis study that the structure will experience localized bending of 1 deg when excited by end forces of 667 lb. Consequently, control forces about the Y axis of 0.11 lb should not affect the system's performance.

A verification of these results was obtained from a digital simulation of the control dynamics of the baseline SSPS structure and control system. Figures 21a through 21f, respectively, show time-history plots of the spacecraft's attitude, control force, and generalized modal velocities, and modal displacements for the first and second anti-symmetrical modes. To determine the influence of structural flexibility, a digital simulation was also performed of the SSPS rigid-body dynamics only. Results are shown in Figures 22a and 22b. A comparison of these two sets of time-history responses reveals that structural flexibility decreases the system's damping ratio and increases its attitude error.

As a result of this flight control performance evaluation about the pitch axis, we concluded that the baseline structure can be considered to be stiffer than necessary for maintenance of the ±1.0 deg pointing accuracy. In response to this observation a parametric study was performed in which attitude control performance sensitivity was measured as a function of changes in structural stiffness.

The results of this study are shown in Figure 23 through 28 in which $\Delta\theta_y/\theta_{cy}$, $\Delta\theta_y/T_{dy}$, T_{R_y}, F_{cy}/T_{dy}, K_{A_y} and K_{R_y} are, respectively, shown as a function of the frequency of the system's lowest vertical anti-symmetrical bending mode frequency. These figures show the results for damping ratios ranging from 0.25 to 1.0 and also consider undamping control frequencies which are factors of 5 and 10 less than the corresponding fundamental bending mode frequency. As expected these results indicate that the system attitude error and response time increases as the structural stiffness (frequency) decreases.

Figure 29 consolidates these pitch mode results by relating structural frequency (stiffness) and attitude error to structural weight. As indicated, a decrease in structural frequency (stiffness) results in a decrease in structural weight, but an increase in attitude error. However, although a 25% change in frequency (stiffness) decreases the structural weight by 50%, the pitch-axis attitude error is still well within the baseline system requirements.

b. Roll Mode

SSPS roll dynamics have been defined to be rotation about the spacecraft's X axis in the Y-Z plane as indicated in Figure 30.

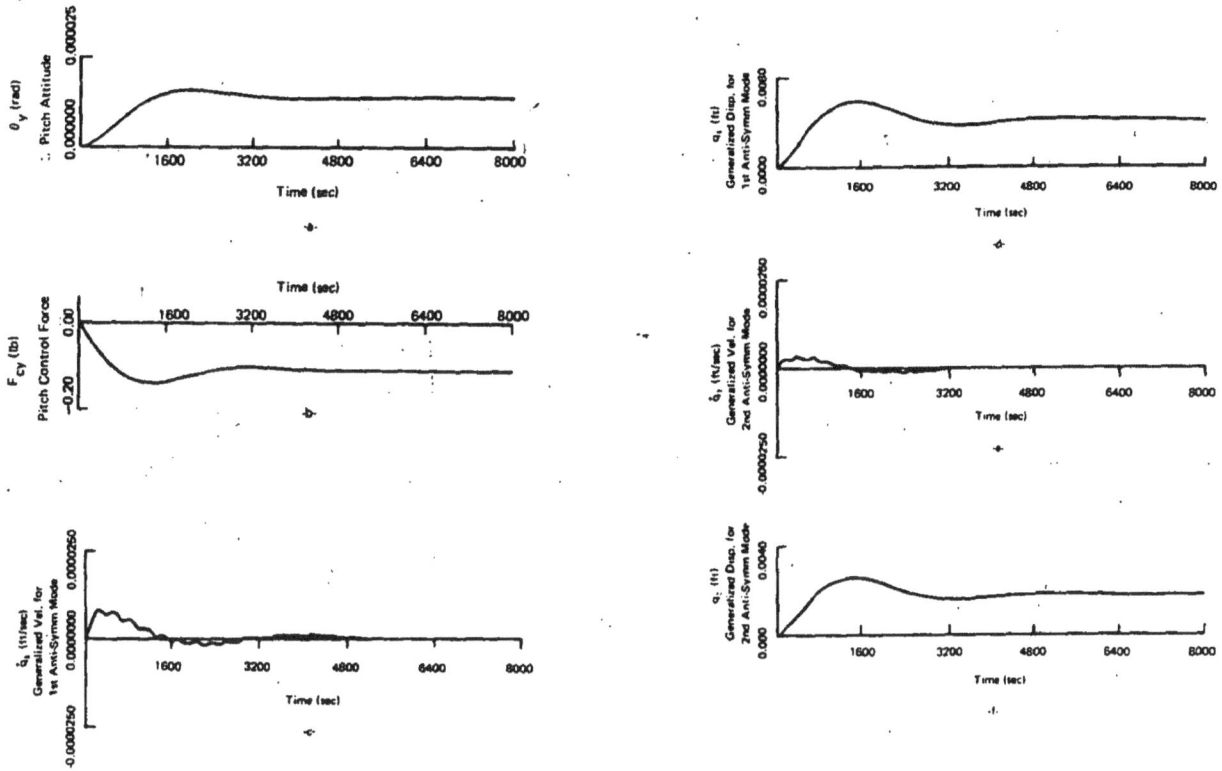

FIGURE 21. — DIGITAL SIMULATION OF PITCH AXIS FLEXIBLE BODY DYNAMICS
FOR A 4100 FT.-LB. CONSTANT DISTURBANCE TORQUE

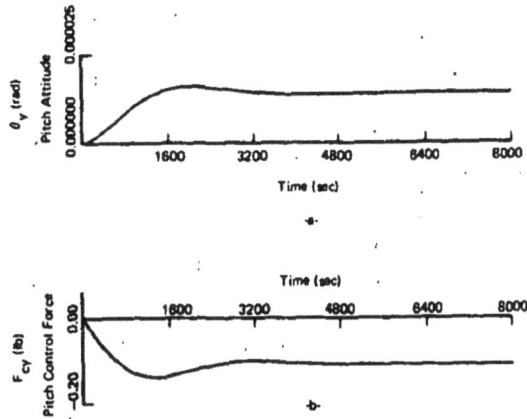

FIGURE 22. — DIGITAL SIMULATION OF PITCH-AXIS, RIGID-BODY DYNAMICS
FOR A 4100 FT.-LB. CONSTANT DISTURBANCE TORQUE

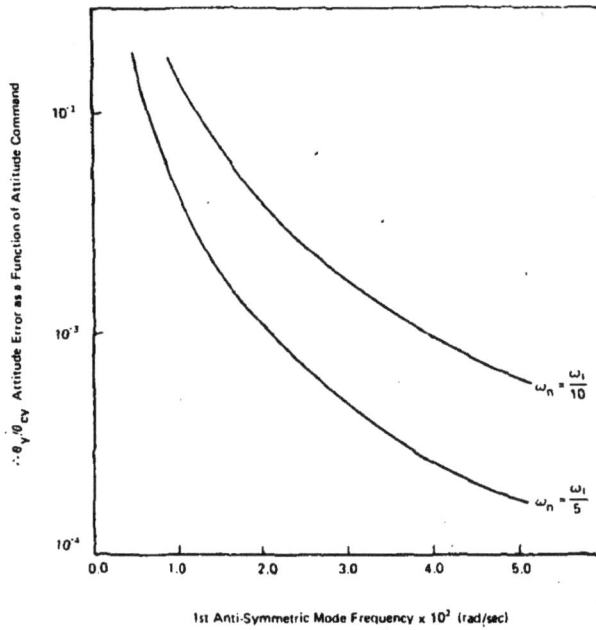

FIGURE 23. — VARIATION OF THE 1st ANTI-SYMMETRIC PITCH MODE FREQUENCY WITH ATTITUDE ERROR

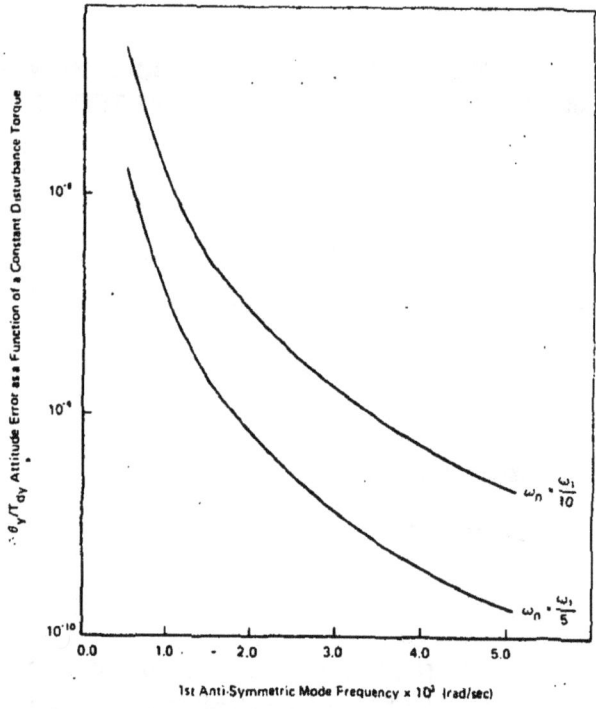

FIGURE 24. — VARIATION OF THE 1st ANTI-SYMMETRIC PITCH MODE FREQUENCY WITH ATTITUDE ERROR

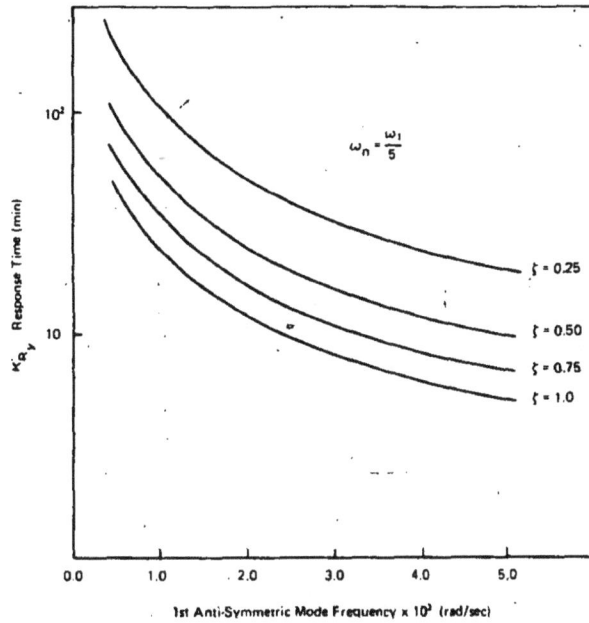

FIGURE 25a. – VARIATION OF THE 1st ANTI-SYMMETRIC PITCH MODE FREQUENCY WITH RESPONSE TIME

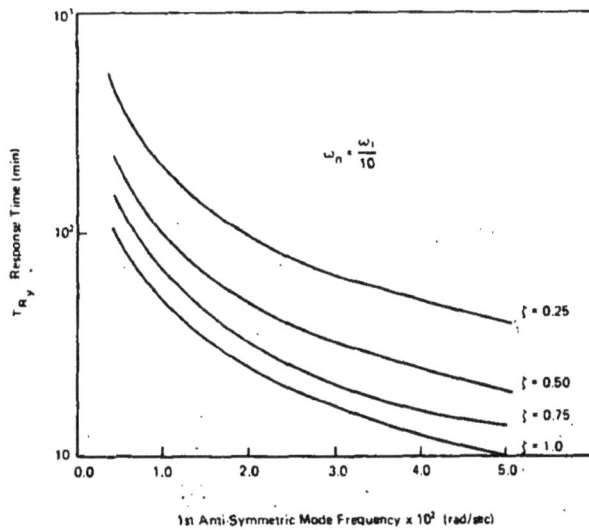

FIGURE 25b. – VARIATION OF THE 1st ANTI-SYMMETRIC PITCH MODE FREQUENCY WITH RESPONSE TIME

48

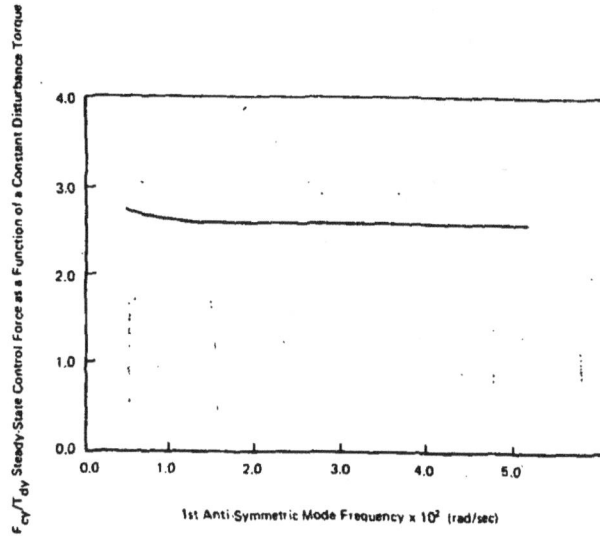

FIGURE 26. – VARIATION OF THE 1st ANTI-SYMMETRIC PITCH MODE FREQUENCY WITH STEADY-STATE CONTROL FORCE

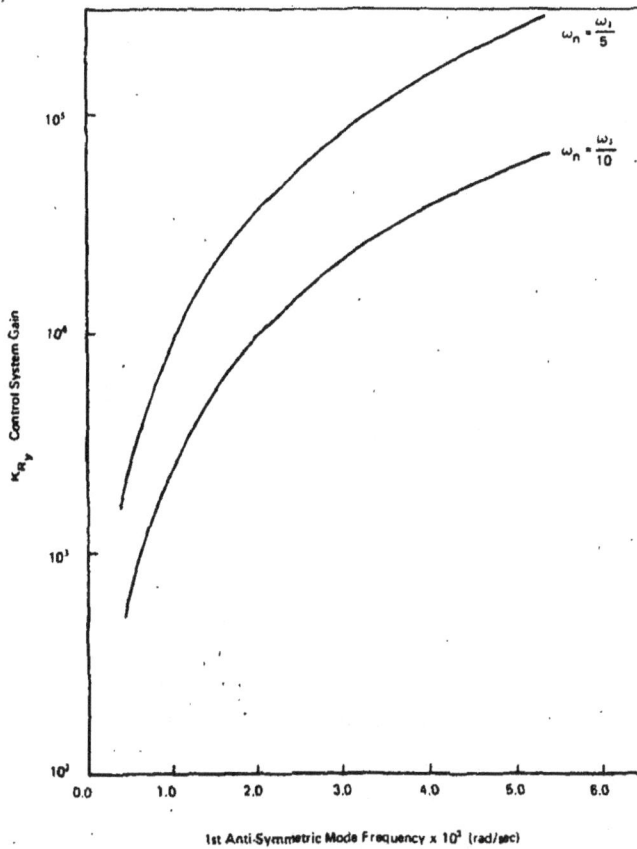

FIGURE 27. – VARIATION OF THE 1st ANTI-SYMMETRIC PITCH MODE FREQUENCY WITH CONTROL SYSTEM GAIN

49

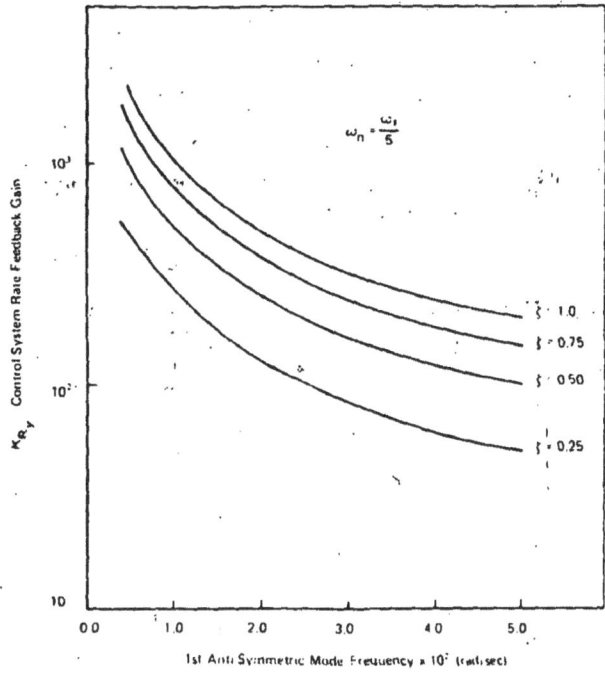

FIGURE 28a. — VARIATION OF THE 1st ANTI-SYMMETRIC PITCH MODE FREQUENCY WITH RATE FEEDBACK GAIN

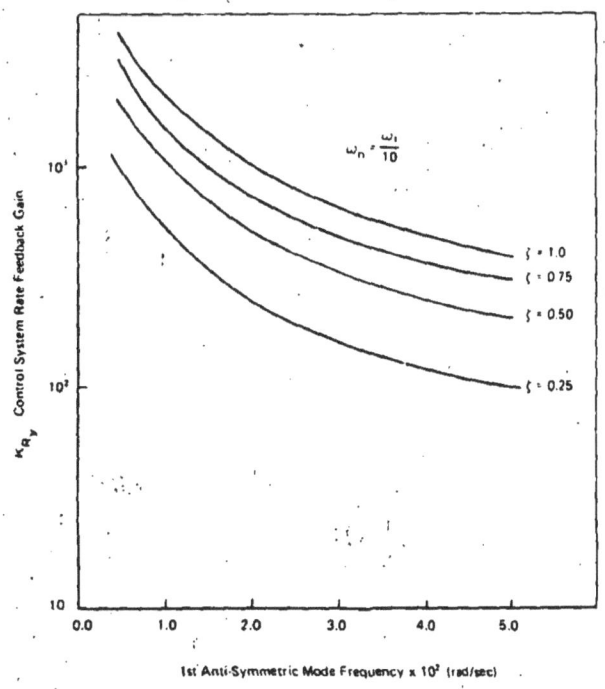

FIGURE 28b. — VARIATION OF THE 1st ANTI-SYMMETRIC PITCH MODE FREQUENCY WITH RATE FEEDBACK GAIN

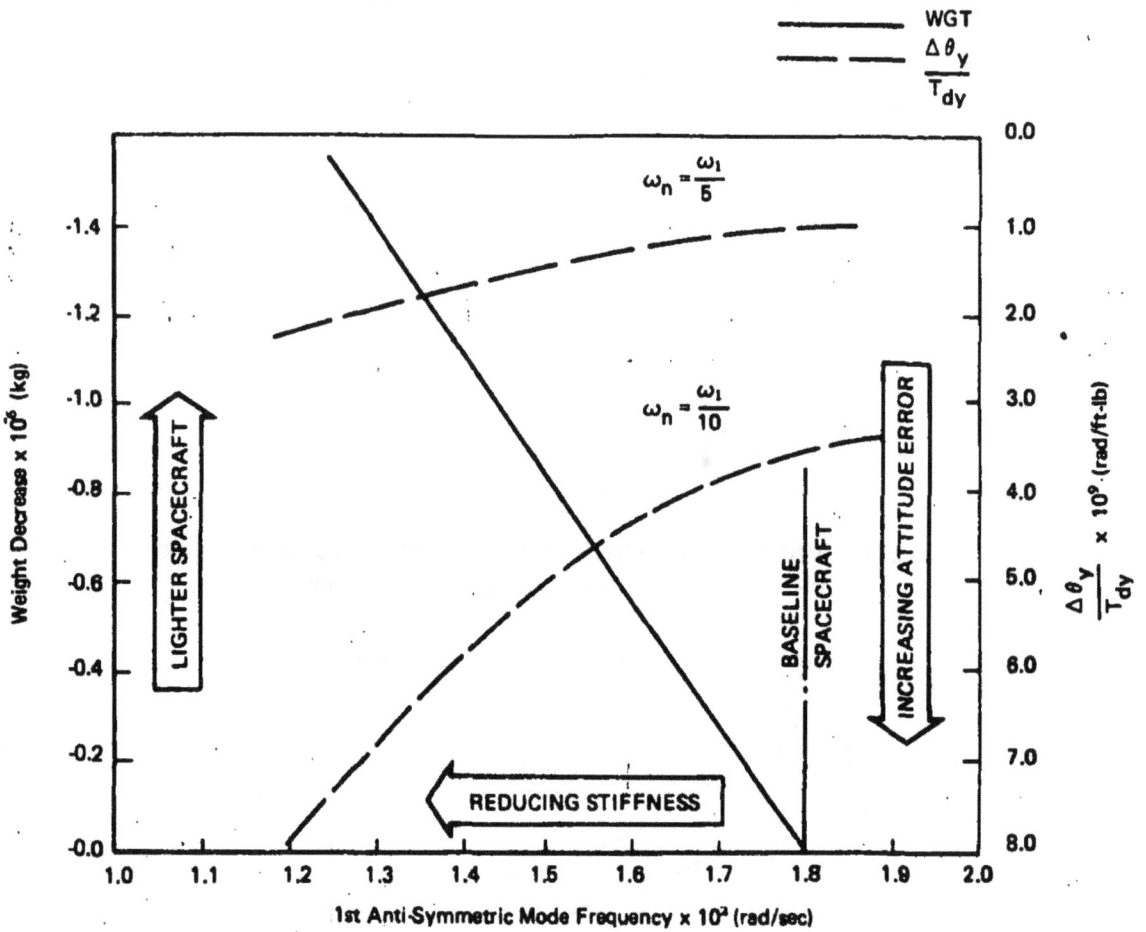

FIGURE 29. — VARIATION OF THE 1st ANTI-SYMMETRIC PITCH MODE FREQUENCY WITH STRUCTURAL WEIGHT AND ATTITUDE ERROR

51

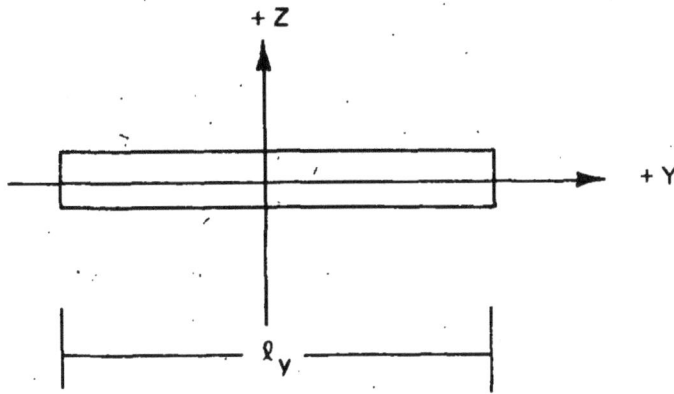

FIGURE 30 ROLL AXIS

The baseline rigid body mass properties of the SSPS about the X axis are listed in Table 9.

TABLE 9

ROLL AXIS MASS PROPERTIES

I_{xx}	Length
14.24×10^6 kg-km²	4.95 km
10.48×10^{12} slug-ft²	3.08 mi.

The structural dynamic analysis about the spacecraft's X axis identified two roll-axis bending modes that would fit the requirements of the math model discussed in Reference 35. The mode shapes of both of these modes are given in Figures 31 and 32. As indicated, the mode shape in Figure 31 is symmetrical, while Figure 32 represents an anti-symmetrical mode.

Table 10 gives modal characteristics for these modes at the actuator and sensor locations located at the axial extremities. A detailed discussion of the roll-axis bending modes is given in Reference 36.

With the SSPS in the nominal orientation with the spacecraft X axis perpendicular to the plane of the synchronous orbit [Reference 35], the SSPS experiences a constant gravity gradient disturbance torque, T_{dx} as well as a solar pressure torque $K_{dx}\theta_x$ which is proportional to the offset rotational angle θ_x about the X axis.

These torques were evaluated in Reference 31 and are repeated in Table 11.

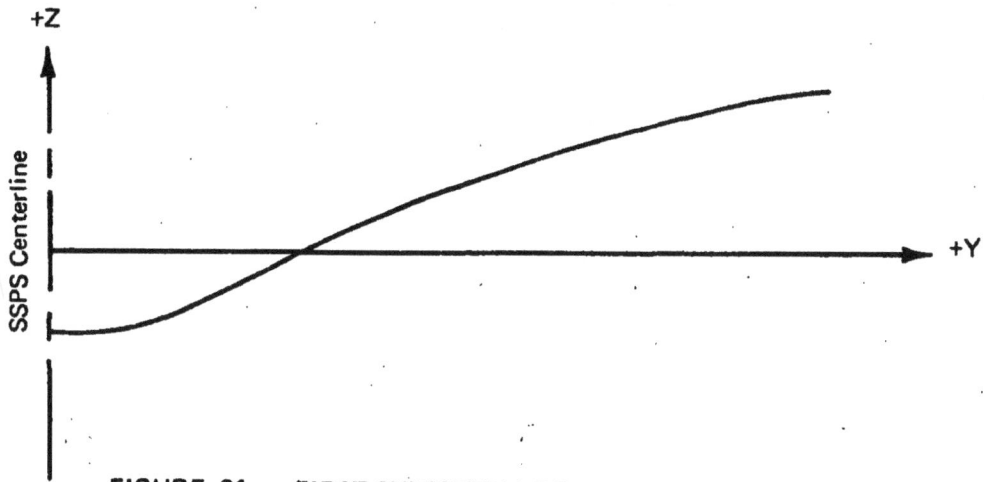

FIGURE 31.— FIRST SYMMETRIC ROLL AXIS BENDING SHAPE

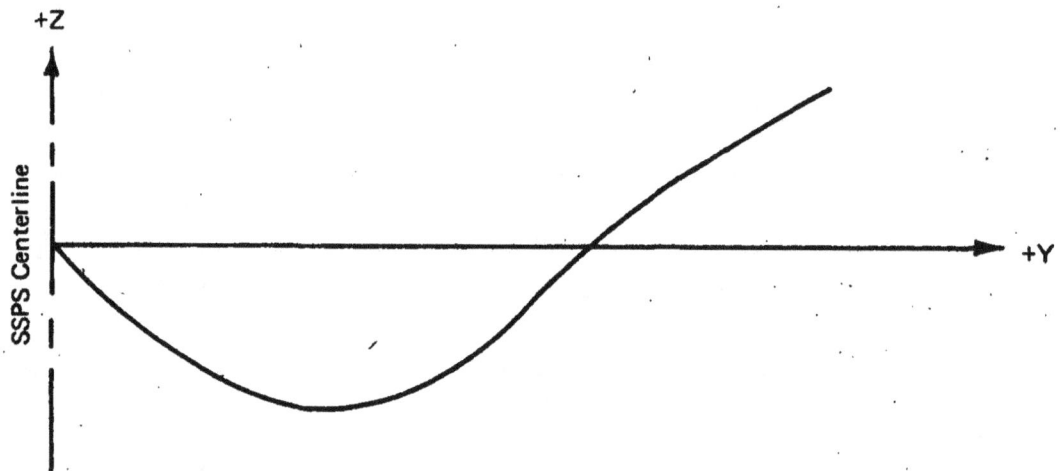

FIGURE 32. — FIRST ANTI-SYMMETRIC ROLL AXIS BENDING MODE SHAPE

TABLE 10

ROLL AXIS BENDING MODE DATA

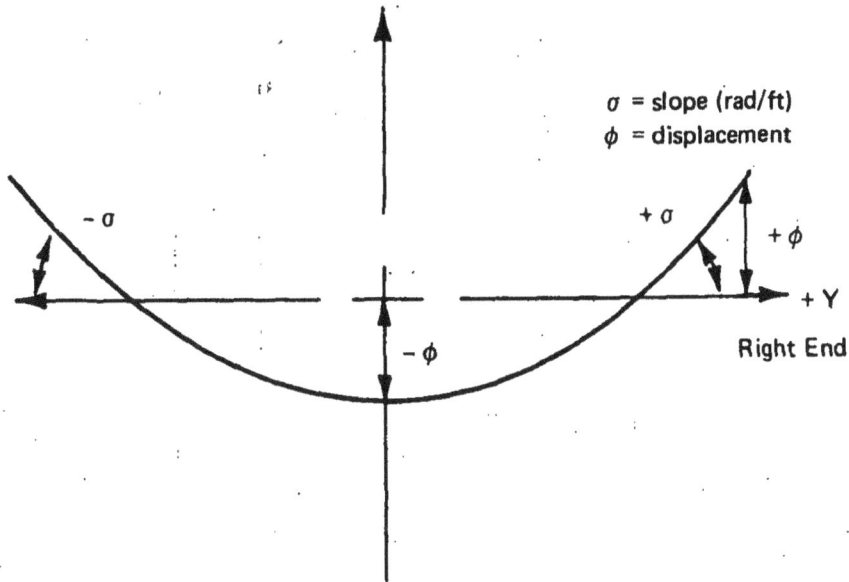

σ = slope (rad/ft)
ϕ = displacement

$-\sigma$

$+\sigma$

$+\phi$

$+Y$

$-\phi$

Right End

Mode	1st SYMM	1st ANTI-SYMM
Frequency (rad/sec.)	0.0453	0.0492
Generalized Mass (slugs)	85008.0	120828.0
Structural Damping	0.0	0.0
Right End Normalized Displacement ϕ	+1.0	+1.0
Right End Normalized Slope σ (rad/ft.)	$+0.568 \times 10^{-3}$	$+0.413 \times 10^{-3}$

54

TABLE 11
ROLL AXIS DISTURBANCE TORQUES

T_{dx}	K_{dx}
12.1 x 10⁴ N-M	7750 N-M/ rad
8.97 x 10⁴ ft -lb	5730 ft-lb/ rad

As discussed earlier, the transient performance requirements for the roll axis called for a damping ratio of 0.5 and an undamped natural frequency equal to a factor of 10 less than the lowest roll-axis anti-symmetrical bending mode frequency. From the model data given in Table 10, this corresponds to an undamped natural frequency of 0.00492 rad/sec.

Using Equations (1) and (2), the roll-axis control parameters which satisfy these performance requirements were found to be:

$$K_{A_x} = 16000.0$$
$$K_{R_x} = 200.0$$

With the above set of baseline parameter values, the dynamic performance of the roll axis was evaluated using the parametric performance indices defined in Tables 1 and 5 in Reference 35. The results of this performance analysis are summarized below:

$$\frac{\Delta\theta_x}{\theta_{cx}} = 2.5 \times 10^{-5} \text{ rad/rad}$$

$$\frac{\Delta\theta_x}{T_{dx}} = 4.2 \times 10^{-9} \text{ rad/ft - lb}$$

$$\frac{F_{cx}}{T_{dx}} = 6.1 \times 10^{-5} \text{ lb/ft - lb}$$

$$T_{R_x} = 1200 \text{ sec}$$

The above results indicate that for each degree that the vehicle is commanded to point from its nominal zero position about the X axis the attitude error for the baseline spacecraft is 2.5×10^{-5} deg. However, this quantity should nominally be zero, since the nominal orientation of the spacecraft requires $\theta_{cx} = 0$. The second term listed about indicates that, in the presence of disturbance torques T_{dx} the system's roll-axis pointing error would be 4.2×10^{-9} rad/ft-lb. Consequently, for a constant-gravity gradient torque of 89,700.0 ft-lb, the roll-axis, steady-state pointing error would be 0.00037 rad (0.0212 deg). To counter this constant disturbance torque, the above relationships indicate that a steady-state control thrust of 5.5 lb is required at each end of the roll-axis acting in opposite directions.

When compared to the ±1 deg baseline system pointing requirements for the roll-axis, it is evident that the presently designed baseline structure and control system are well within the performance specifications. Furthermore, a control thrust of 5.5 lb is far below the maximum 667-lb limit placed on the end thruster by the results of the structural analysis [Reference 36].

The results of this flight control performance evaluation were verified using a digital simulation of the control dynamics of the SSPS. Figures 33a through 33e show time-history response plots for the spacecraft's roll attitude, control thrust profile, and generalized bending mode dynamics in the presence of a 89,700-ft-lb disturbance torque. A similar set of time-history plots is shown in Figure 34 for the roll-axis, rigid-body dynamics. A comparison of these sets of dynamic responses shows that the influence of structural flexibility is to decrease the system's damping ratio and increase the attitude error.

On the basis of this flight control performance analysis about the roll axis, it was concluded that the baseline structure and control system were well within the design specifications. Consequently, a parametric study was performed to determine the influence of structural stiffness upon the system's attitude error. The resulting set of curves are presented in Figures 35 through 40 in which the frequency of the lowest anti-symmetric roll bending mode is, respectively, plotted against $\Delta\theta_x/\theta_{cx}$, $\Delta\theta_x/T_{dx}$ T_{R_y}, F_{cx}/T_{dx}, K_{A_y}, and K_{R_y}.

As noted, these plots consider a family of systems damping ratios ranging from 0.25 to 1.0 as well as two sets of undamped natural frequencies corresponding to values which are factors of 5 and 10 below the lowest anti-symmetrical bending modes of the roll axis.

The results of this roll mode parametric study indicate that the system's roll axis attitude error and response time increase as the roll axis fundamental bending frequency (stiffness) decreases. Figure 41 cross-plots structural weight and roll attitude error against bending mode frequency. The results shown in this plot indicate that a 25% decrease in the baseline structural frequency results in a 50% reduction in structural weight, while the baseline pointing accuracy is still well within required accuracy.

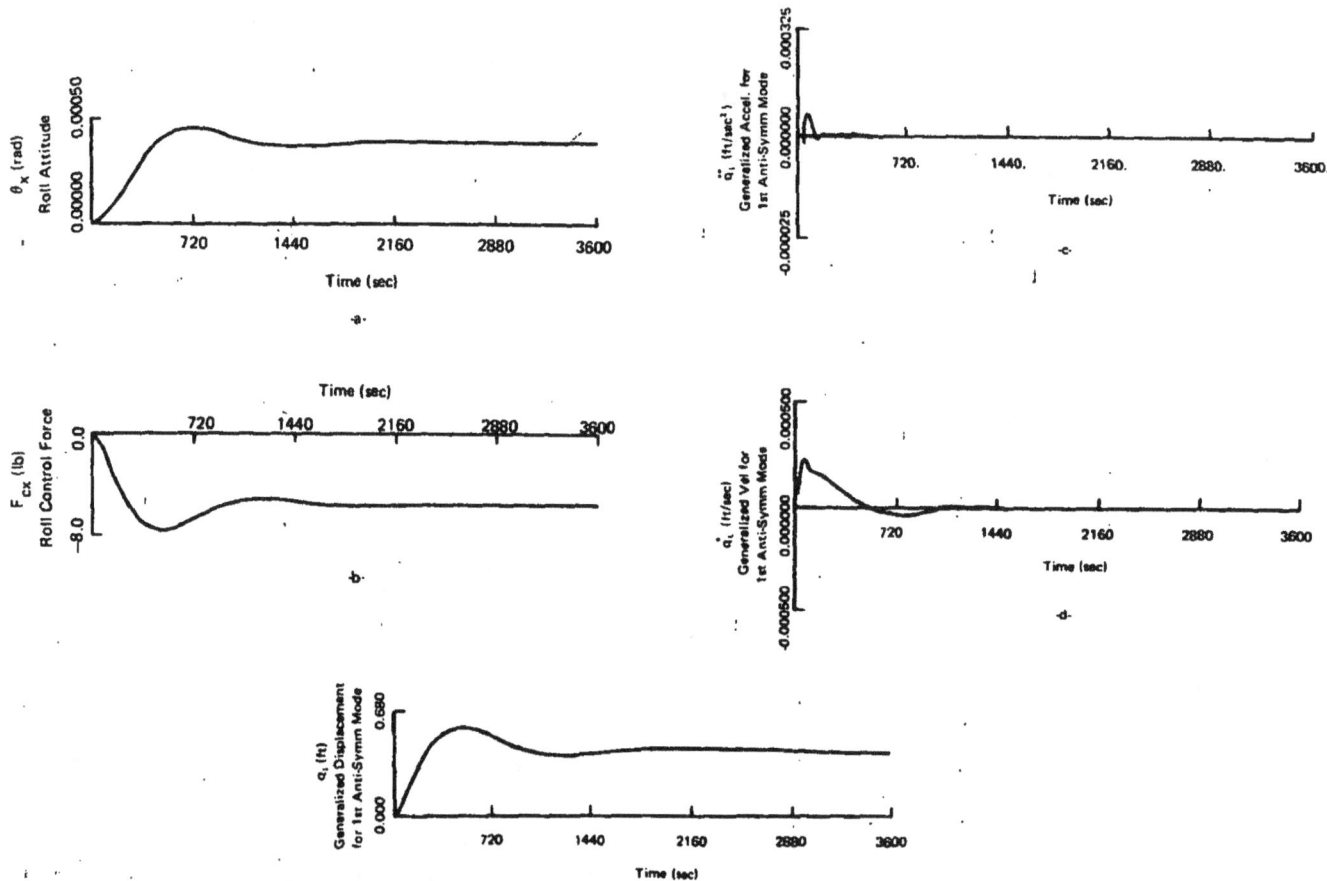

FIGURE 33. — DIGITAL SIMULATION OF ROLL AXIS FLEXIBLE BODY DYNAMICS
FOR A 89,700 FT.-LB. CONSTANT DISTURBANCE TORQUE

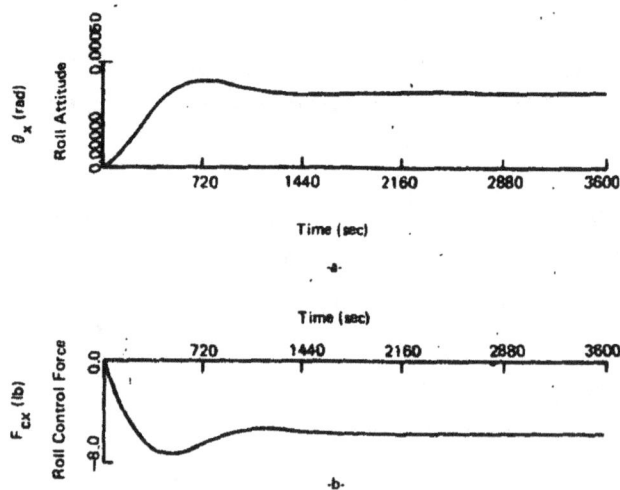

FIGURE 34. — DIGITAL SIMULATION OF ROLL-AXIS, RIGID-BODY DYNAMICS
FOR A 89,700 FT.-LB. CONSTANT DISTURBANCE TORQUE

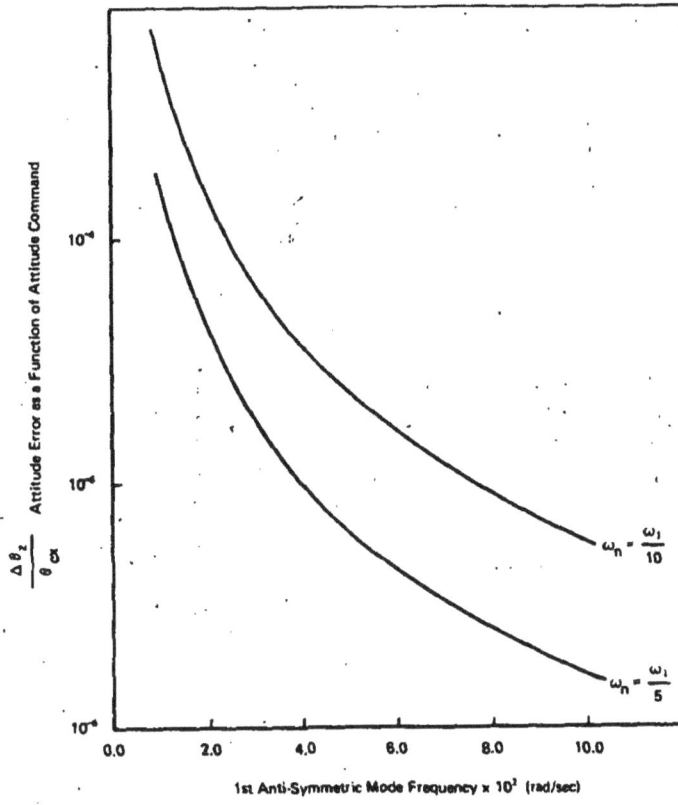

FIGURE 35. — VARIATION OF THE 1st ANTI-SYMMETRIC ROLL
MODE FREQUENCY WITH ATTITUDE ERROR

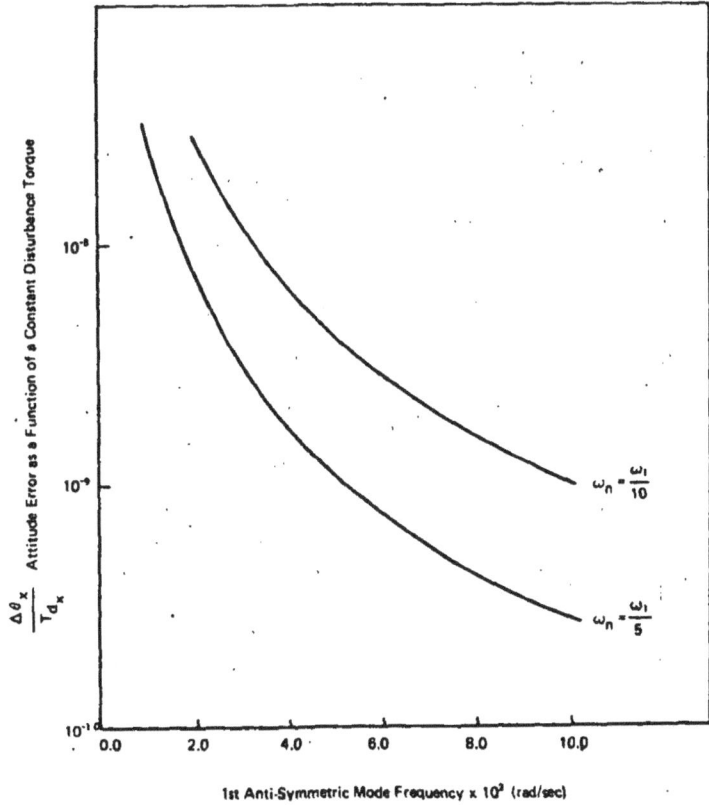

FIGURE 36. — VARIATION OF THE 1st ANTI-SYMMETRIC ROLL
MODE FREQUENCY WITH ATTITUDE ERROR

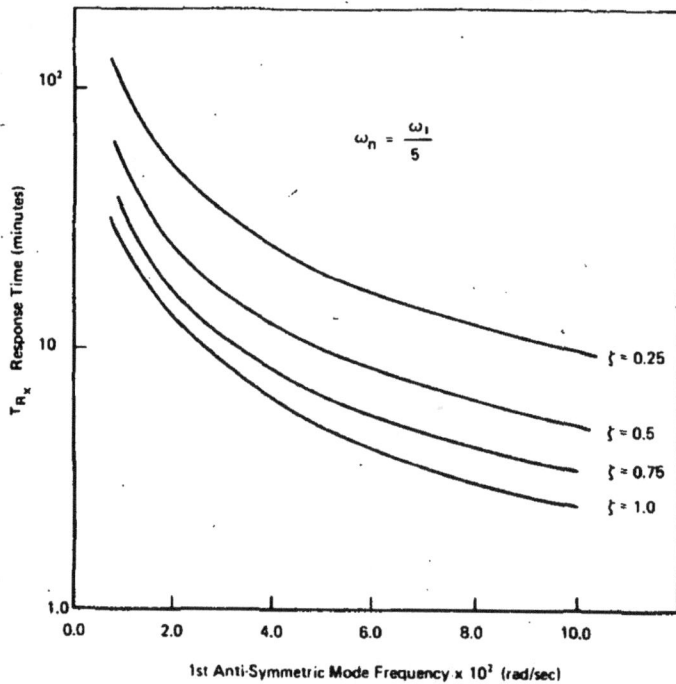

FIGURE 37a. – VARIATION OF THE 1st ANTI-SYMMETRIC ROLL MODE FREQUENCY WITH RESPONSE TIME

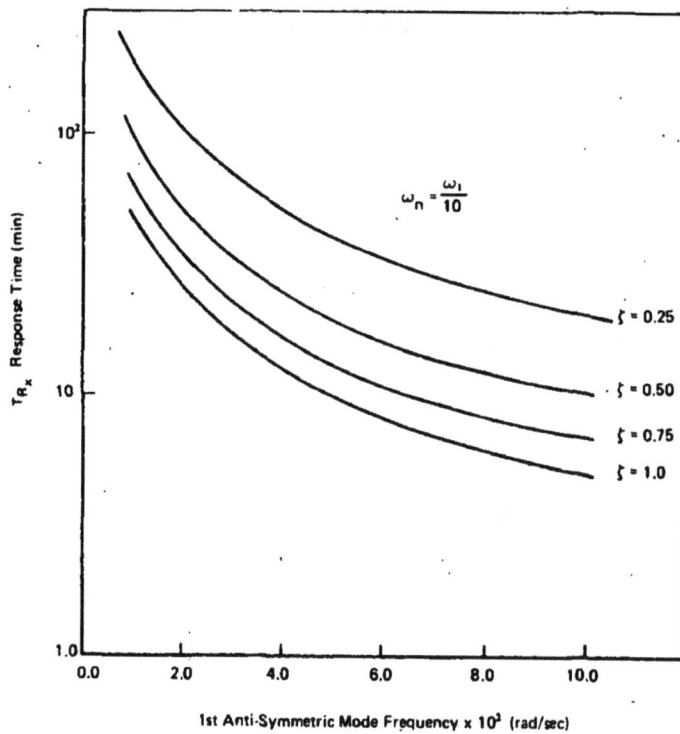

FIGURE 37b. – VARIATION OF THE 1st ANTI-SYMMETRIC ROLL MODE FREQUENCY WITH RESPONSE TIME

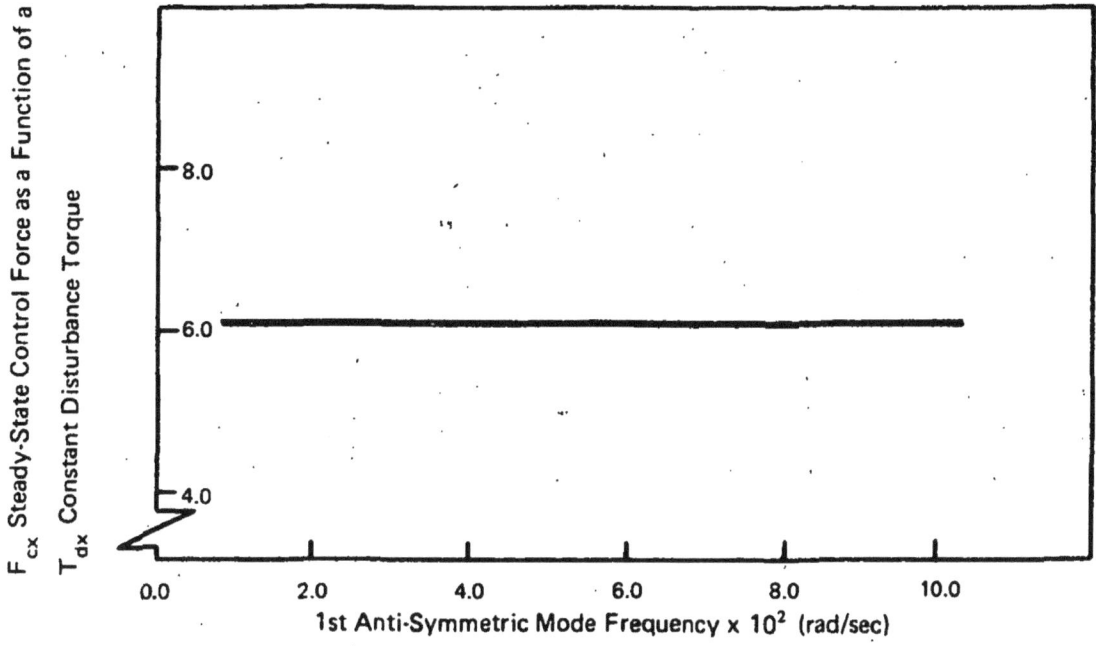

FIGURE 38. — VARIATION OF THE 1st ANTI-SYMMETRIC MODE
FREQUENCY WITH STEADY-STATE CONTROL FORCE

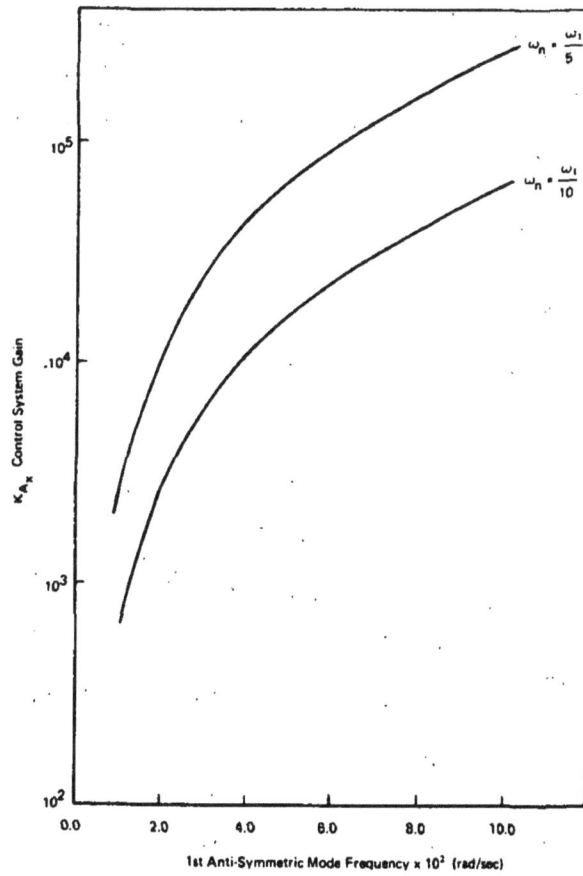

FIGURE 39. — VARIATION OF THE 1st ANTI-SYMMETRIC ROLL
FREQUENCY WITH CONTROL SYSTEM GAIN

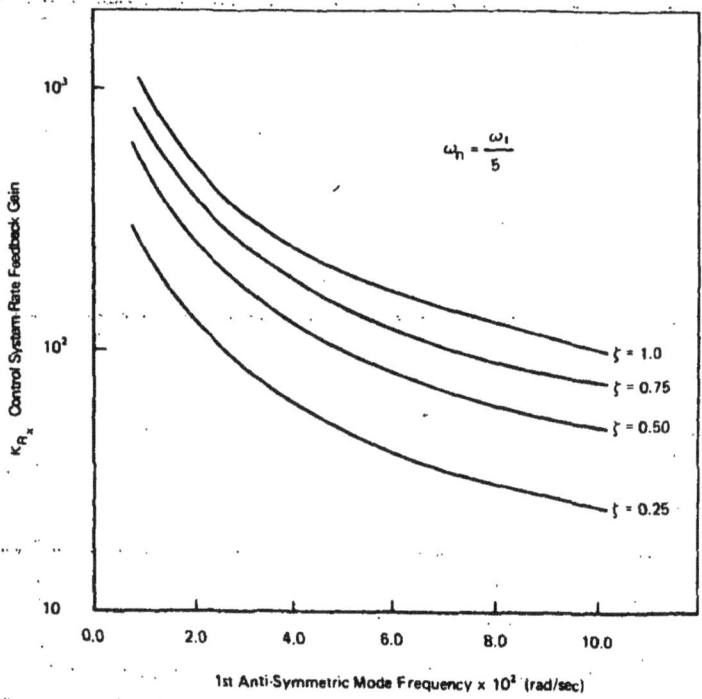

FIGURE 40a. – VARIATION OF THE 1st ANTI-SYMMETRIC ROLL MODE WITH RATE FEEDBACK GAIN

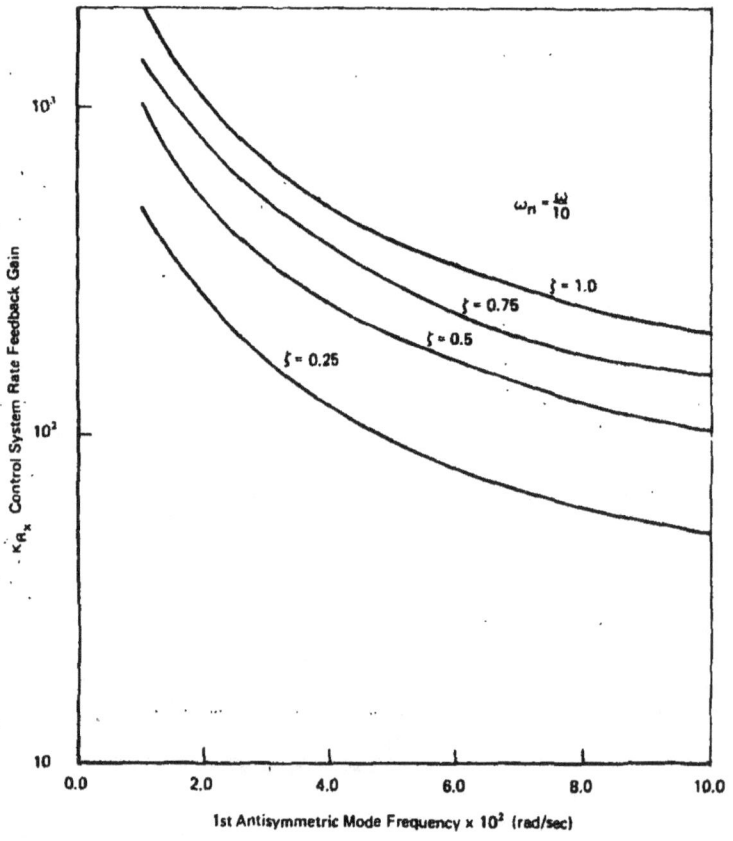

FIGURE 40b. – VARIATION OF THE 1st ANTI-SYMMETRIC ROLL MODE WITH RATE FEEDBACK GAIN

61

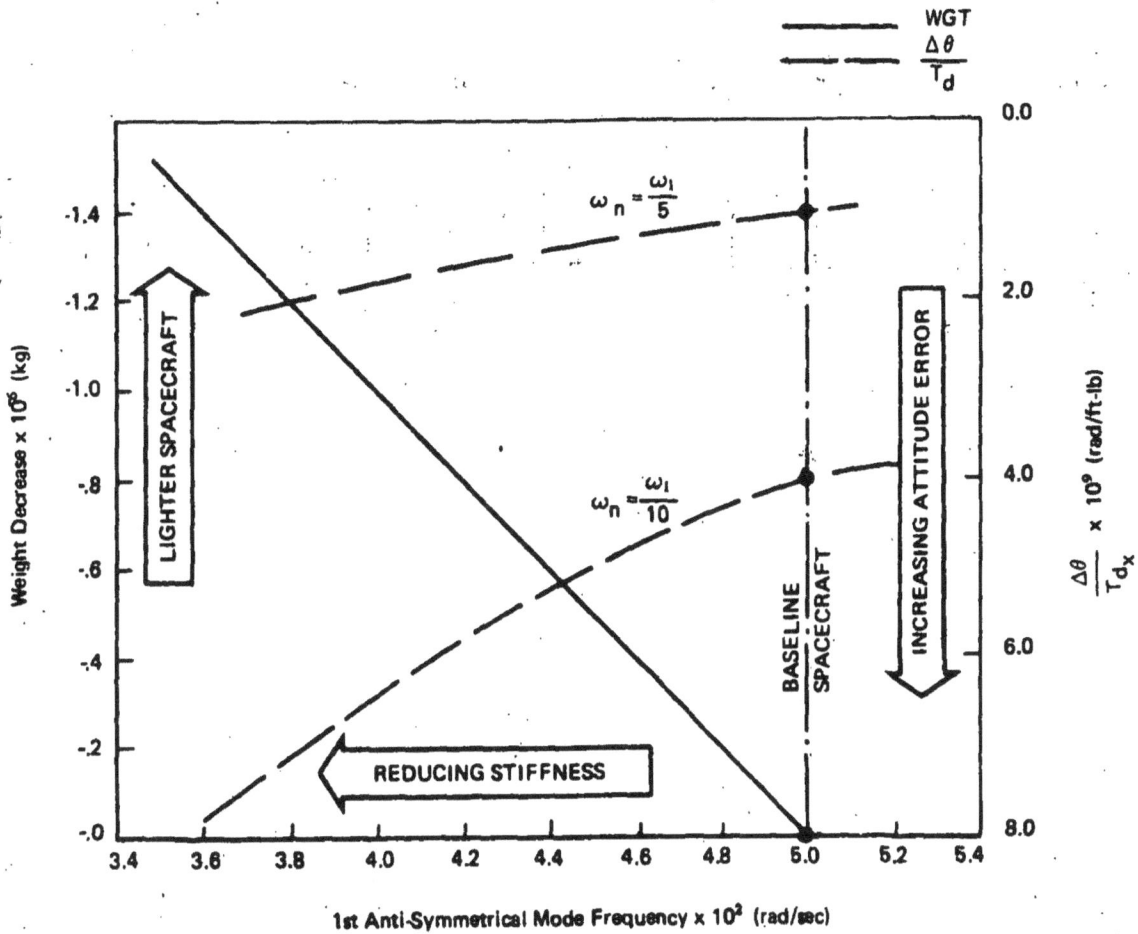

FIGURE 41. – VARIATION OF THE 1st ANTI-SYMMETRIC ROLL MODE FREQUENCY
WITH STRUCTURAL WEIGHT AND ATTITUDE ERROR

c. Yaw Mode

Yaw mode dynamics has been defined to be rotation about the spacecraft's Z axis in the X-Y plane as shown in Figure 42.

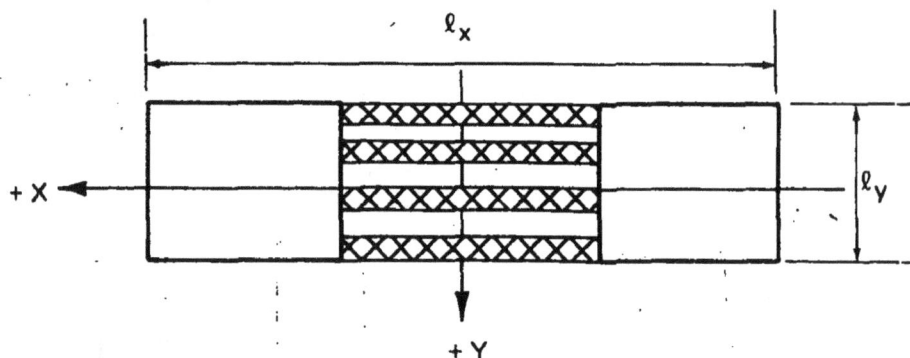

FIGURE 42   YAW AXIS

The baseline rigid body mass properties of the SSPS about the Z axis are listed in Table 12.

TABLE 12

YAW AXIS MASS PROPERTIES

I_{ZZ}	Length
137×10^6 kg-km²	12 km
100.83×10^{12} slug-ft²	7.46 mi.

An analysis of the structural dynamics about this axis identified four lateral bending modes for the baseline structure. The mode shapes for each of these modes are shown in Figures 43 through 46. As indicated, Figures 43 and 45 represent symmetric bending modes, while Figures 44 and 46 represent anti-symmetric modes. The modal data at the extremities of the structure in this yaw mode are given in Table 13. Data are presented for only one side, since the structure is assumed to be symmetrical. A detailed discussion of these mode shapes is given in Reference 36.

In its nominal orientation the spacecraft's axis is perpendicular to the orbit plane and the Z-axis oscillates in the X-Z plane about the Y-axis at 23-1/2 deg/year. In this orientation the SSPS only experiences a solar pressure torque which is proportional to the angular deviations of the spacecraft about the Z-axis. Numerical values for these torques are given in Table 14.

For the baseline spacecraft it was found that in order for the yaw-mode dynamics to be characterized by a 0.5 damping ratio and an undamped natural frequency of 0.0025 rad/sec 1/10 lowest anti-symmetric yaw bending mode, it is necessary that the attitude control have the following control gains (from Equations (1) and (2)).

$$K_{A_z} = 16000.0$$

$$K_{R_z} = 400.0$$

63

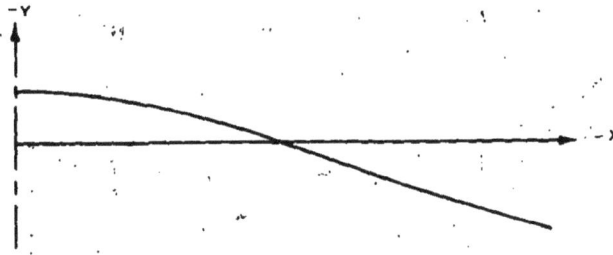

FIGURE 43. — 1ST SYMMETRIC YAW-AXIS BENDING MODE

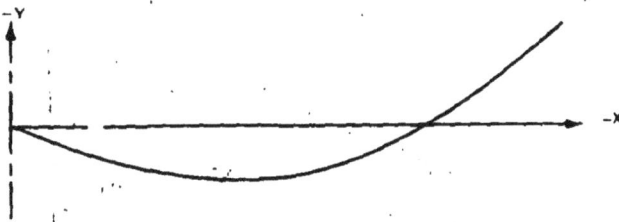

FIGURE 44. — 1st ANTI-SYMMETRIC YAW-AXIS BENDING MODE

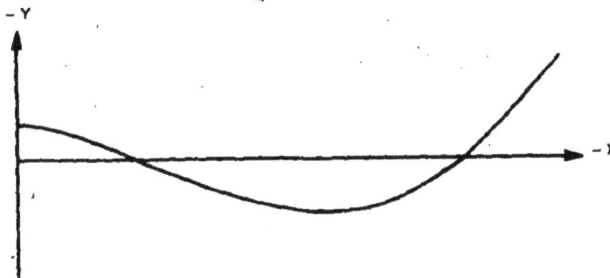

FIGURE 45. — 2nd SYMMETRIC YAW-AXIS BENDING MODE

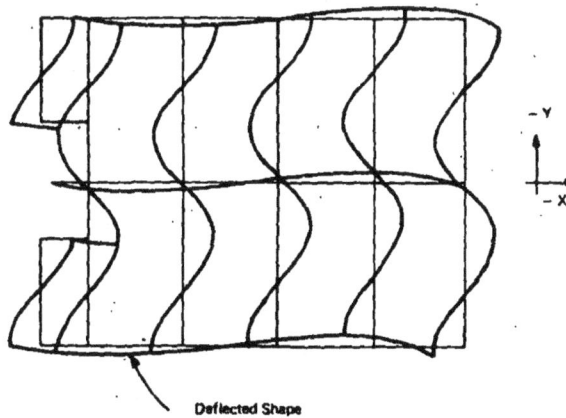

Deflected Shape

FIGURE 46. — 2nd ANTI-SYMMETRIC YAW-AXIS BENDING MODE

64

TABLE 13
YAW AXIS BENDING MODE DATA

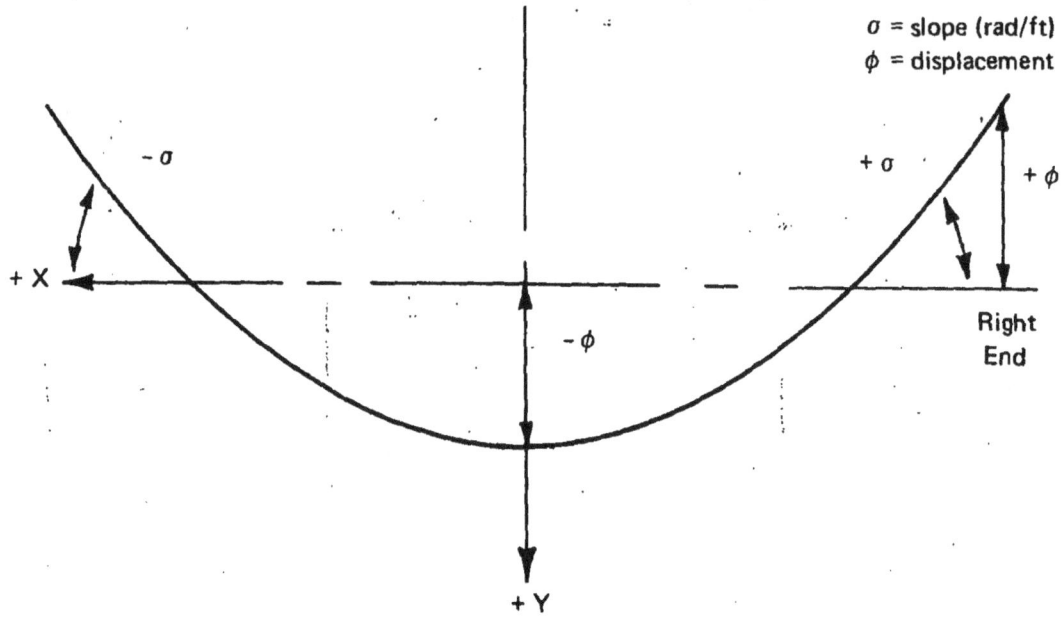

σ = slope (rad/ft)
ϕ = displacement

Mode	1st SYMM	1st ANTI-SYMM	2nd SYMM	2nd ANTI-SYMM
Frequency (Rad/Sec)	0.01093	0.025	0.0319	0.0451
Generalized Mass (Slugs)	207864.	157200.	150480.	277152.0
Structural Damping	0	0	0	0
Right End Normalized Displacement, ϕ	- .989	+ .992	+ .994	+ .246
Right End Normalized Slope σ (Rad/Ft)	- .089x10^{-3}	+ .214x10^{-3}	+ .188x10^{-3}	+ .13x10^{-3}

TABLE 14

YAW AXIS DISTURBANCE TORQUES

T_{dz}	K_{dz}
0	155 x 10^4 ft-lb/rad
	114.6 x 10^4 ft-lb/rad

With the above set of baseline parameter values the yaw-axis dynamic performance was evaluated using the parametric performance indices defined in Tables 1 and 5 of References 35 and 42. The results of this analysis are summarized below.

$$\frac{\Delta\theta_z}{\theta_{cz}} = 2.0 \times 10^{-3}$$

$$\frac{\Delta\theta_z}{T_{dz}} = 1.65 \times 10^{-9} \text{ rad/ft-lb}$$

$$\frac{F_{cz}}{T_{dz}} = 2.7 \times 10^{-5} \text{ lb/ft-lb}$$

$$T_{R_z} = 2200 \text{ sec}$$

Since the spacecraft's nominal orientation requires θ_{cz} to be equal to zero, and since in this orientation the spacecraft experiences zero constant disturbance torques about the Z axis, the yaw-axis attitude error is zero. However, a digital simulation of the initial condition response for the yaw mode of the SSPS was performed to identify its fundamental characteristics. The time-history plots for the axis flexible body and rigid body dynamics are shown in Figures 47a through 47b and Figures 48 a through 48b, respectively.

Although this analysis indicates that the SSPS will not be disturbed about the Z-axis from its nominal orientation, a parametric study was performed to investigate the sensitivity of the baseline design to variations in structural stiffness. The results of this parametric study are shown in Figures 49 through 56.

A comparison of these figures with Figures 23 through 28 indicates the similarity between the yaw-mode and pitch-mode characteristics. This was an expected result due to the fact that their respective mass properties are nearly identical.

The results of the yaw mode parametric study are summarized in Figure 56 in which structural weight and attitude error are cross-plotted against structural frequency. As expected, structure weight decreases with frequency, while attitude error and response time increase.

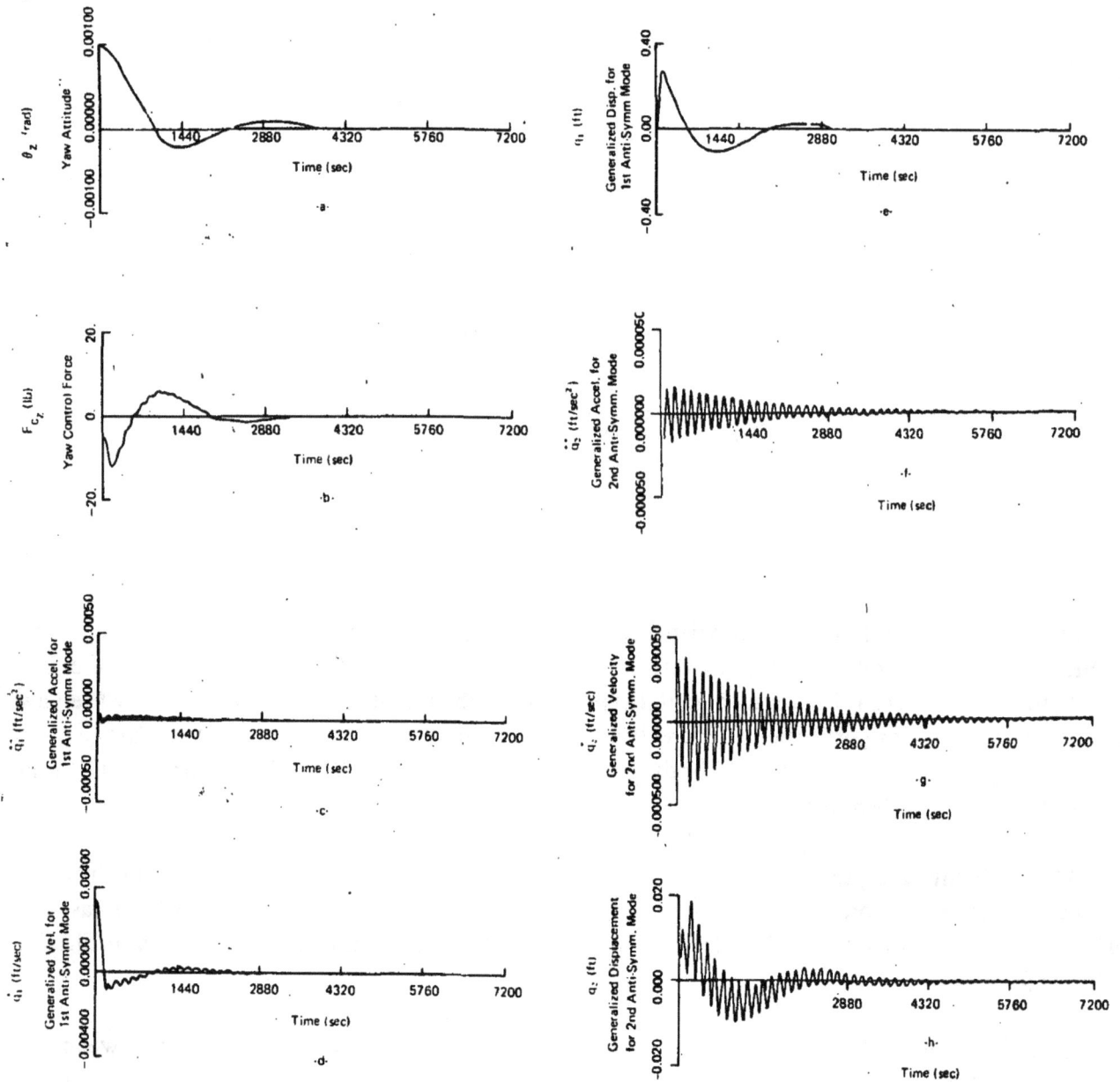

FIGURE 47. — DIGITAL SIMULATION OF THE FLEXIBLE BODY DYNAMICS
FOR A 0.001 RAD-INITIAL ATTITUDE

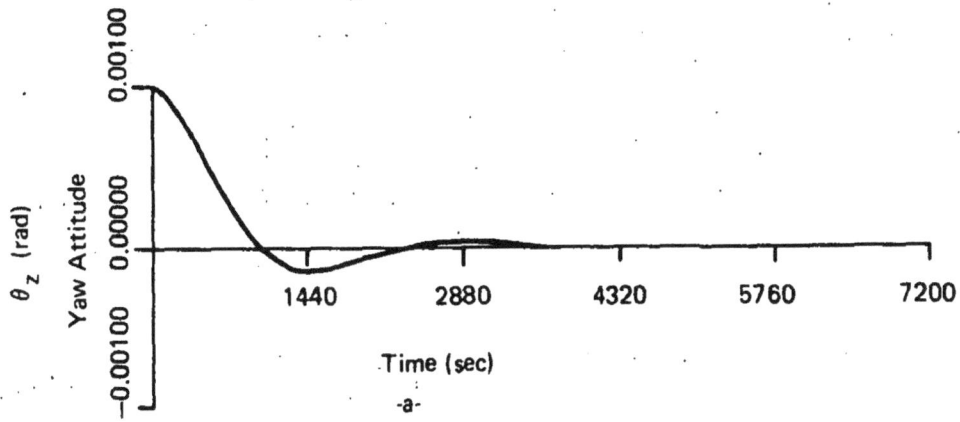

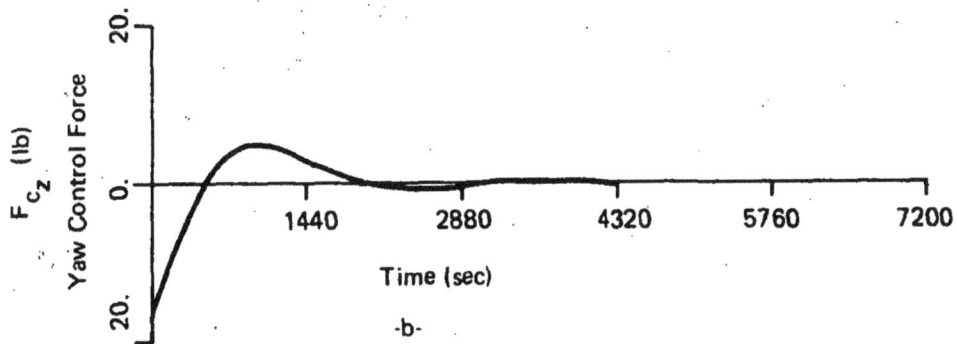

FIGURE 48. — DIGITAL SIMULATION OF THE YAW AXIS RIGID BODY DYNAMICS
FOR A 0.001-RAD-INITIAL ATTITUDE

68

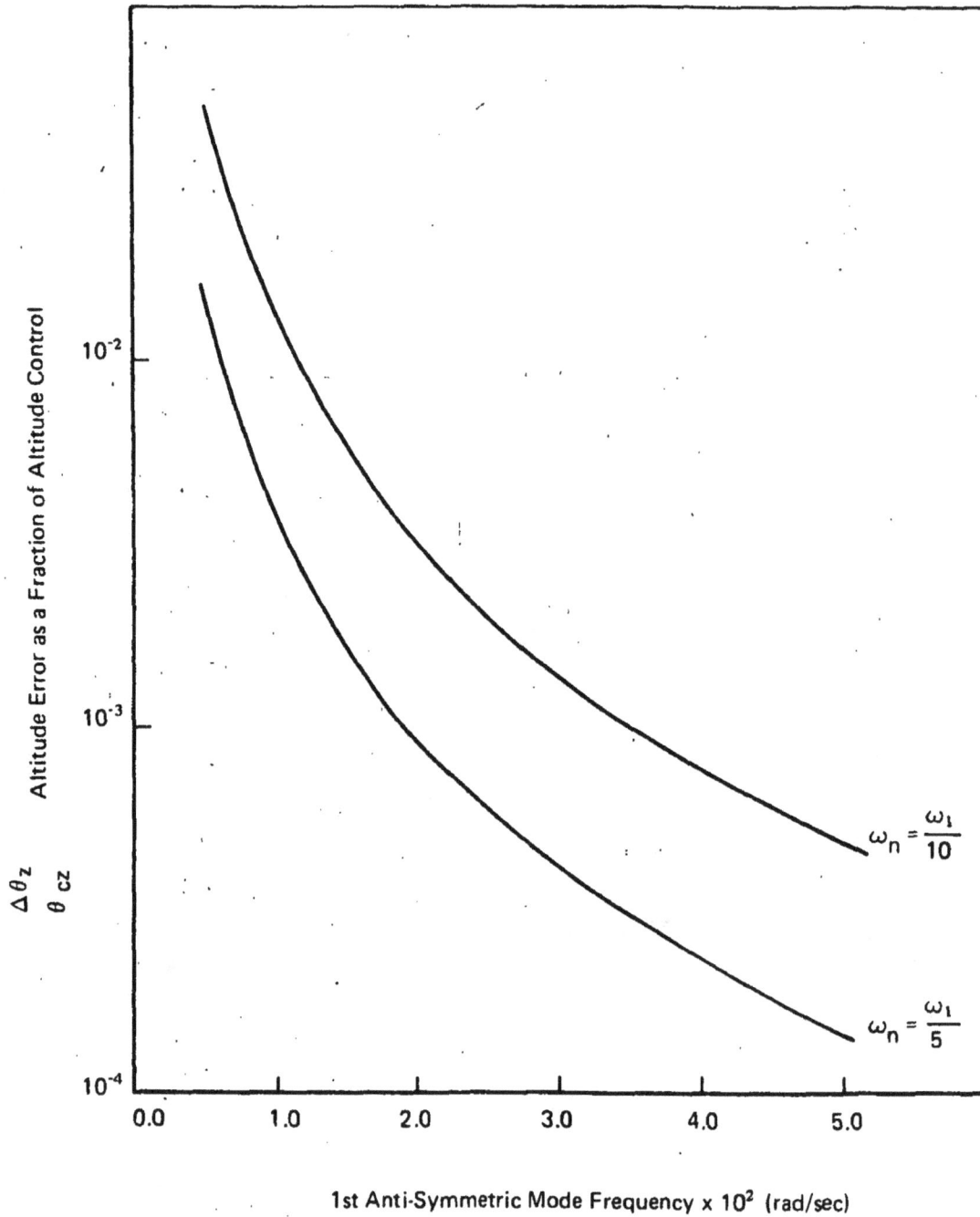

FIGURE 49. — VARIATION OF THE 1st ANTI-SYMMETRIC YAW MODE
FREQUENCY WITH ATTITUDE ERROR

69

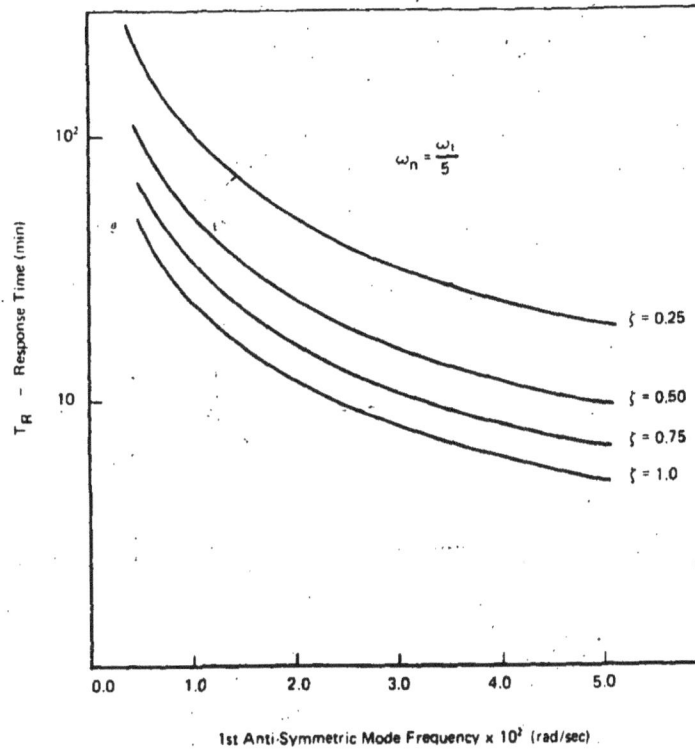

FIGURE 50. — VARIATION OF THE 1st ANTI-SYMMETRIC YAW MODE
FREQUENCY WITH RESPONSE TIME

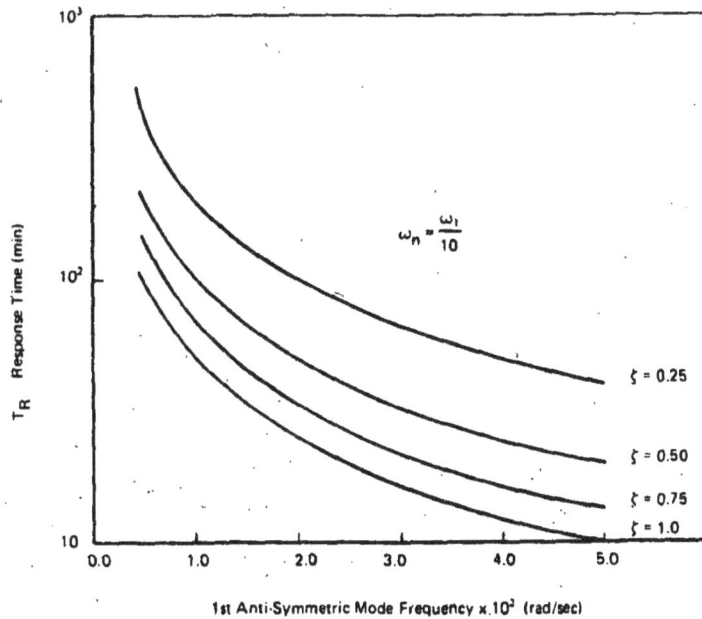

FIGURE 51. — VARIATION OF THE 1st ANTI-SYMMETRIC YAW MODE
FREQUENCY WITH RESPONSE TIME

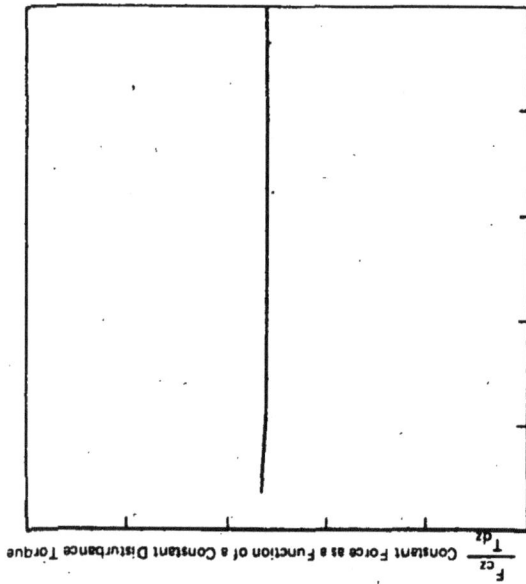

FIGURE 53. — VARIATION OF THE 1st ANTI-SYMMETRIC YAW MOD
FREQUENCY WITH CONTROL SYSTEM GAIN

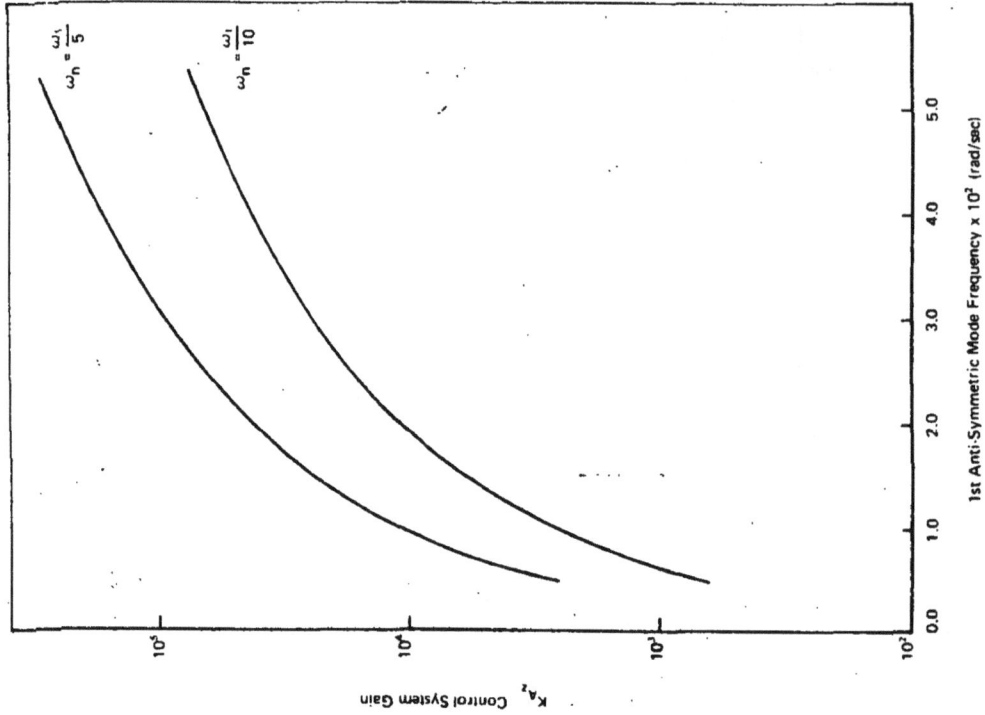

FIGURE 52. — VARIATION OF THE 1st ANTI-SYMMETRIC YAW MODE
FREQUENCY WITH CONTROL FORCE

71

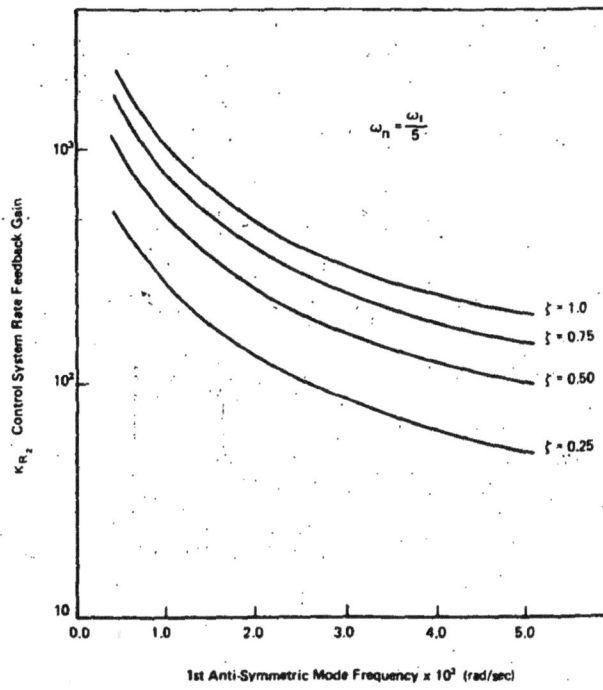

FIGURE 54. — VARIATION OF THE 1st ANTI-SYMMETRIC YAW MODE
FREQUENCY WITH RATE FEEDBACK GAIN

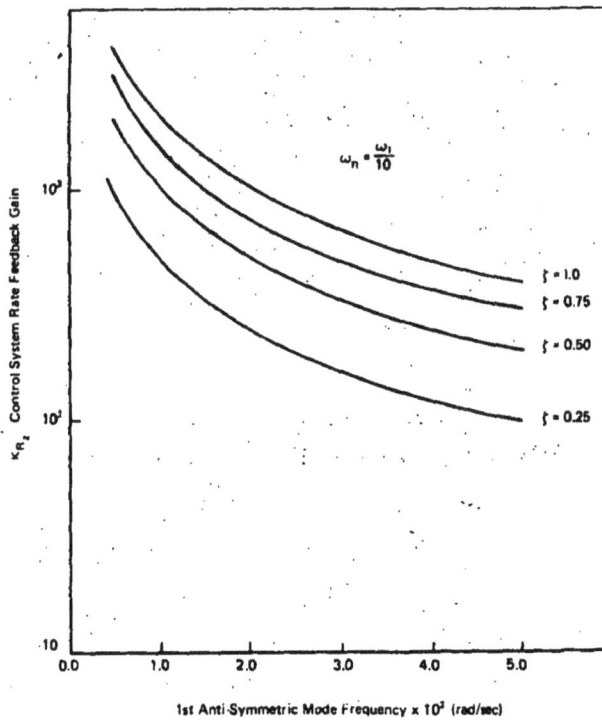

FIGURE 55. — VARIATION OF THE 1st ANTI-SYMMETRIC YAW MODE
FREQUENCY WITH RATE FEEDBACK GAIN

72

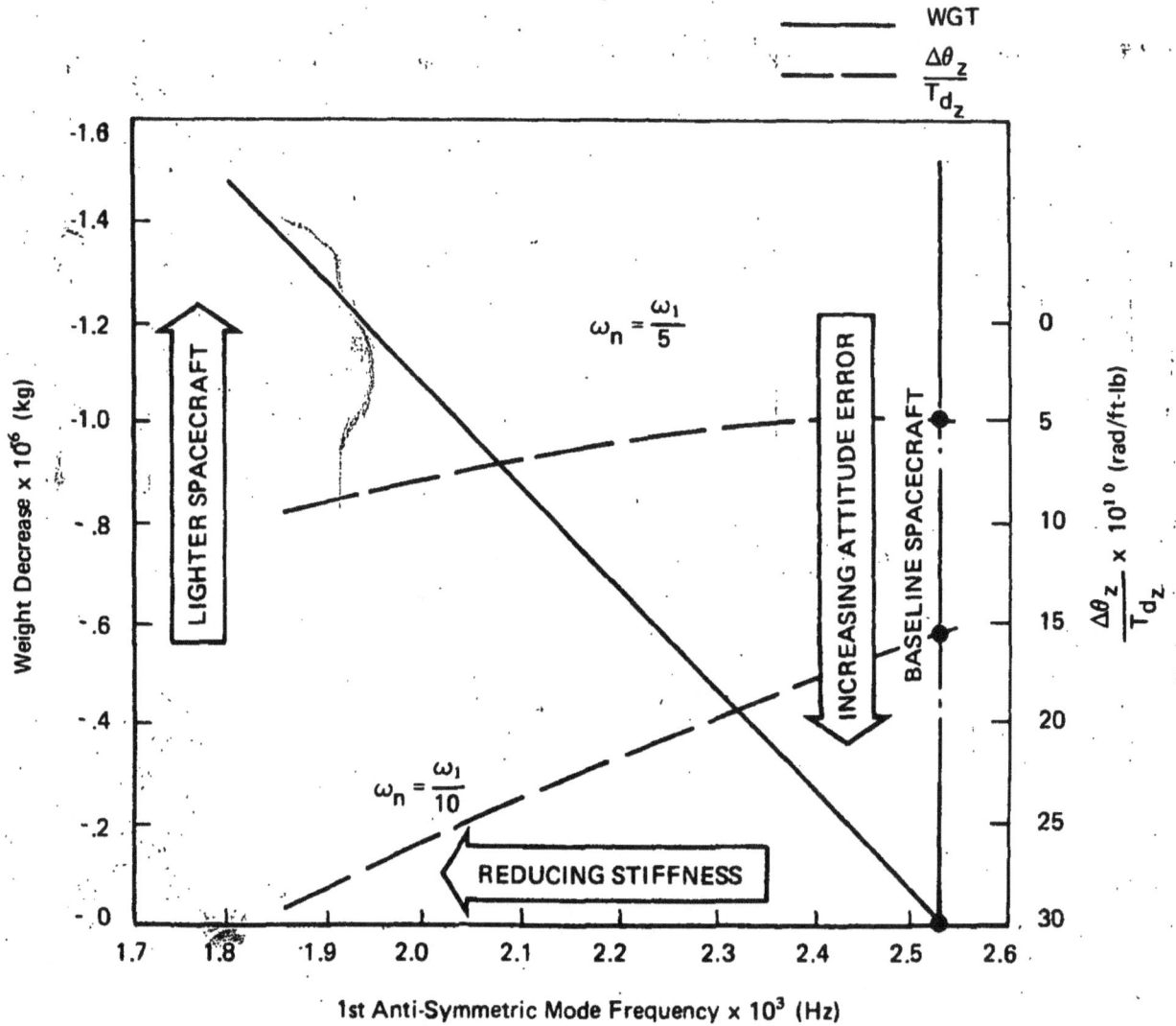

FIGURE 56. — VARIATION OF THE 1st ANTI-SYMMETRIC YAW MODE FREQUENCY
WITH STRUCTURAL WEIGHT AND ATTITUDE ERROR

73

Although this study indicates that the SSPS would not be disturbed about the Z axis from its nominal orientation, it is realized that in a practical sense this will not be the case. Therefore, it is suggested that future structure and control interaction studies be expanded to include the effect of cross-coupling and torsional bending, as well as a system of non-linearities such as control system dead zones and saturation limits.

Identification of Areas Requiring Further Flight Control Performance Analysis. – The results of the structural analysis [Reference 43] indicate the predominance of torsional vibrational modes, as well as cross-coupling between all modes. However, the math model presently employed in the controllability study cannot accept torsional modes and is restricted to the analysis of only uncoupled modes. While this model has proven adequate for this feasibility study, it is suggested that for future studies the math model be expanded from a planar model to a three-dimensional model which can accept all the bending modes identified in Reference 43. Furthermore, it is felt that an analysis of this expanded system could only be accomplished through the use of a non-real time simulation. A digital computer program has been developed at Grumman under contract #NAS 10 10973 which would fulfill the above requirements. It is felt that this program would provide a good interim capability for predicting the three-dimensional dynamic behavior of an SSPS. This time-history program can treat rotating or non-rotating satellites of any shape or mass distribution. The satellite may be idealized by using up to 100 masses. Since the rotatory inertia of each mass is considered, up to 600 physical coordinates may be employed. A maximum of 20 elastic modes are then used to represent the elastic motions of the vehicle and to reduce the number of coordinates being integrated. Control systems may be added to this program in subroutine form. The program has already been used to analyze the behavior of a realistically modeled space station containing five different types of control systems.

Before using this program, the gravity-gradient and orbital centrifugal loads would be computed as a function of time by assuming that the vehicle is rigid and in perfect control. These loads would then be used as the time-dependent forcing function for the program and the dynamic response of the vehicle would be obtained. Since the gravity-gradient and orbital centrifugal loads are a function of the vehicle attitude, the true loads might differ from the computed loads. Thus, it may be necessary to recompute the loads based on the program's output and repeat the time-history run using the new loads. One or more iterations of this procedure may be required.

Since gravity-gradient and orbital-centrifugal loads are primary disturbances, for the final analysis another program should be developed which is similar to the above program, but has the capability to include these effects more directly and accurately. The loads would be determined including the vehicle's orientation (response), and an iterative approach would no longer be required. The baseline attitude control system presently used for the SSPS is linearly modeled and uses proportional thrust actuators. It is recommended that, in the expanded study, the control system be modified to include such non-linearities as dead zone characteristic and control-thrust saturation limits. Although pointing accuracy has been found to be well with the ±1 deg pointing requirements, the actual attitude control system could conceivably have an attitude dead zone of ±1 deg.

Therefore, it is recommended that the controllability studies be expanded before steps are taken to resize the baseline system.

Conclusions.— The flight control performance evaluation of the baseline system and its nominal structural design found the SSPS to be well within a ±1 deg attitude control specification about all three major axes. The results of parametric studies indicated that, although attitude error performance decreases with structural stiffness, a 50% decrease in structural weight (and commensurate reduction in structural stiffness) will still allow the SSPS to remain well within attitude control requirements about all three axes.

On the basis of these results it was concluded that structurally the baseline SSPS is overdesigned about all three axes. However, as suggested, additional studies should be conducted before a structural redesign is undertaken.

RFI AVOIDANCE TECHNIQUES

Optimization of Microwave Transmission System

Background. – The feasibility assessment of the microwave generation, transmission, and rectification for the SSPS was carried out in support of the baseline configuration definition. The substantial output of microwave energy requires that the radio frequency interference (RFI) effects of an SSPS also be assessed.

Discussions with personnel of the Office of Telecommunications Policy have set the guidelines for the RFI assessment; however, this investigation was limited in scope to cover only what is currently believed to be the most likely part of the microwave band of frequencies. Should the SSPS be found to have the capacity to supply a significant portion of future power needs, it would enjoy a priority in frequency allocation such that the required bandwidth at the near optimum frequency could be made available and alternative approaches identified to compensate the displaced users.

It is possible to select a near optimum range for the fundamental frequency from 2 to 4 GHz. Above this range, confidence in the design of devices, such as the Amplitron and associated filtering, decreases and the attenuation of the microwave beam by water-laden cloud increases, particularly for the low-probability instantaneous meteorological events. From the SSPS point of view alone, it would be preferable on the one hand to tend toward the high end to reduce antenna size requirements and, on the other hand, to tend toward the low end where wavelengths are longer and precision in manufacturing of such elements as filters are not as critical. However, at the lower frequencies there is high impact on existing users in many countries.

The conduct of definitive trade-off studies to indicate a near optimum frequency selection requires detailed design of the devices, including filters, and estimates of associated costs. Such detailed design investigations will be required to assure that the design goals for noise and harmonic filtering can be met.

Since a near-optimum frequency for the SSPS from a design point of view could be identified anywhere in the 2- to 4-GHz band, the selection of a fundamental frequency within this band should be made from the point of view of minimizing interference with other users.

To carry out the RFI task a model and a set of assumptions for the microwave transmission system (37) were defined. The model included orbital and ground location, ground power transmission, device characteristics, phase-front control, efficiencies, induced RF environment, ionospheric and atmospheric attenuation, major frequency segment, specific frequency, typical users, and selected equipment.

Based on a set of assumptions for filter design and recognizing existing allocated radio astronomy and fixed satellite space-to-Earth bands, a frequency of 3.3 GHz was selected as the fundamental frequency for the main power beam of the SSPS and further RFI investigations were conducted on this basis.

RFI due to continuous operation of the microwave beam has been examined in this task. Transients associated with start-up, shut-down, and physical interference due to clouds, aircraft, and the onset of shadowing by the Earth will have to be investigated in future studies. Data, control, and command links will have to be included in future investigations of RFI and associated frequency allocation.

The phase front was assumed to be controlled by a system similar to that described in Reference 38. It was further assumed that 90% of the power could be collected within the main lobe. This assumption can be realized by relatively simple distribution of power density illumination across the transmitting antenna (see "Effects of SSPS RFI on Other Users" later in this chapter).

Near optimum orbital locations were selected in an equatorial synchronous orbit at 123° W and 57° E Longitude based on reducing propellant requirements for RCS units. Near optimum ground locations were selected in the southwest desert area, 33° N Latitude, 113° 30' W Longitude for further specific investigations.

These investigations concerned the RFI associated with the fundamental power transfer beam, including its side lobes, its noise, spurious signals, and harmonics, and cover primarily the continuous operations phase as compared to start, stop, and failure transients. Their purpose was to delineate the approaches which would reduce RFI and to provide an outline for subsequent studies.

Discussion. --

a. Orbital Location

The SSPS should be located at the stable node in equatorial synchronous orbit (e.g., ~ 123° W Longitude), primarily for station keeping purposes. If a satellite the size of the SSPS should lose its activity ability to station-keep at synchronous altitude and start moving around in orbit, it would sweep out large regions that may be occupied. "Rules of the Road" will undoubtedly be evolved which will dictate that the smaller (spacecraft proportions) most maneuverable spacecraft shall "give way" to the largest, most unmaneuverable spacecraft.

After the first prototype has become operational, additional antennas and solar arrays could be added to an SSPS. A similar set could be located at the second stable node at ~ 57° E Longitude. After sufficient confidence is gained, SSPS's could be located at other points and rely on active station-keeping systems. RFI induced by an SSPS will affect its close neighbors in orbit. This effect is discussed in "Effects of SSPS RFI on Other Users" later in this chapter.

b. Ground Location

The guidelines associated with receiving antenna locations in the vicinity of four selected high power-consuming areas were set up as follows:

1. Favorable meteorological location;
2. Relatively unpopulated region; and
3. Within a few hundered miles of the using city.

A location in the desert in the Southwest was identified, primarily based on its being a relatively unpopulated area from the "man, animal, and bird" points of view.

In the time period 1990 to 2000, when the system would be operational, a power distribution network would exist (e.g., superconducting underground power cables), such that facilities having surplus capabilities for power generation could assist in the supply of the peak need for user areas in parts of the country remote from the power source. Although detailed economic studies may indicate differently, the efficiencies are so very high and the RFI, due to underground power transmission, is so very low that their development in large magnitude appears highly probable. There is not much difference in overall efficiency (less than 3%) for a particular user site in the continental United States to receive power beamed relatively locally compared to its being beamed to a near optimum ground site and then transferred by superconducting ground cable to the user site.

Favorable meteorological locations are not as meaningful to site selection because most "favorable meteorological locations" exhibit high rainfall rates, even for the low probabilities of occurrence and unscheduled "brown outs" are considered sufficiently undesirable to favor the lower frequencies. The 3.3-GHz frequency choice gives low attenuation, even at high precipitation rates, so weather will not be a vital factor in ground location. Ground winds, ice, sand and, for off-shore installations, waves will play a more vital role in site selection when detailed design analyses of ground sites and their equipment are undertaken.

With respect to the "relatively unpopulated region" guideline, it appears now that national and international land utilization studies, based on projected Earth resources data, may well shed different light on the subject. On the one hand, those areas that are not utilized currently, simply because they do not have the requisite natural water, may be highly productive if supplied with water by one of several projected means. On the other hand, the simple existence of an SSPS ground station, depending on how its power sales are regulated, may result in a focusing of industry near or even beneath the site. In-depth study of effects on local bird population will be essential to understand whether birds will avoid or be attracted by the microwave beam, and if they are attracted by its warming effect, will they become an unacceptable nuisance: e.g., a low-temperature ambient environment may be such that the beam attracts; conversely a high-temperature ambient environment may lead to avoidance by birds. These study results may therefore become particularly significant in site-selection criteria.

In summary, ground site selection criteria will be greatly influenced by results of projected Earth resource studies and by social and political considerations. Current and projected bird population at the site could be major factors in site selection. The more narrow SSPS system aspects of site selection lend themselves to relatively simple and known analysis techniques.

Device Design. — The Amplitron is a very desirable microwave generator for the SSPS, because of its inherent characteristics such as high efficiency, light weight, and high reliability. In this section, the Amplitron design versus its operating frequency for the SSPS concept is discussed. The frequency range considered first extended from 1 to 5 GHz. However, in addition to the microwave generation aspect, there are many other reasons for selecting a particular frequency for the SSPS, such as antenna design, the interaction of the microwave beam with the Earth's atmosphere, and noise interference.

Initial studies on the type of Amplitron needed for SSPS application showed that each tube in the phased-array system would have a relatively low-power range of 5 to 10 kW. The Amplitron designed during our preliminary study is shown in Figure 57 and is the basis for the study of Amplitron design versus frequency for the SSPS program. This figure shows the essential components of the Amplitron design from the point of view of heat flow and weight. Some of the features of the SSPS Amplitron include:

- A heavy outside tube envelope which will not be required in the space application;

- Samarium-cobalt magnets which greatly reduce the magnet weight; and

- Pyrolytic graphite which is attached to the Amplitron for radiation cooling.

The basic SSPS Amplitron was designed for a power output of approximately 5 kW at a frequency of 2 GHz and the temperature drop in the vanes was limited to 50°C. A comparison of the vane temperature rise versus frequency is shown in Figure 58. At the higher frequencies, the heat path surface area is decreasing so that the vane temperature rises.

The disposal of waste heat resulting from the energy conversion process will be a major problem in space. However, the high efficiency of the Amplitron will keep the waste heat at a minimum. Passive cooling of microwave generators may be employed by attaching radiating fins made from pyrolytic graphite. This material operating in the temperature range of 250° to 400°C has a heat conductivity nearly twice that of copper. With the pyrolytic body at 325°C, the vane temperature is shown over the 1- to 5-GHz frequency range (see Figure 59).

The high efficiency of the Amplitron will reduce the dissipation of wasted power in the anode circuit and thus allow this circuit to be conduction-cooled. The overall efficiency of the Amplitron can be attributed to its circuit efficiency and internal dc-to-rf conversion efficiency. The internal conversion efficiency is dependent primarily upon the value of the magnetic field utilized, while the circuit efficiency is dependent upon the $I^2 R$ losses. The percentage of the dc input power dissipated in the anode circuit of the SSPS Amplitron versus frequency is shown in Figure 60. Since the circuit loss is dependent upon the skin depth, there is a variation of anode losses with frequency.

Iron Flux Return Path

SmCo Magnet

Cooling Radiator (Anode)

Anode

Cathode

Cooling Radiator (Cathode)

Vanes of Amplitron

Scale: 1 Inch

SECTION A – A

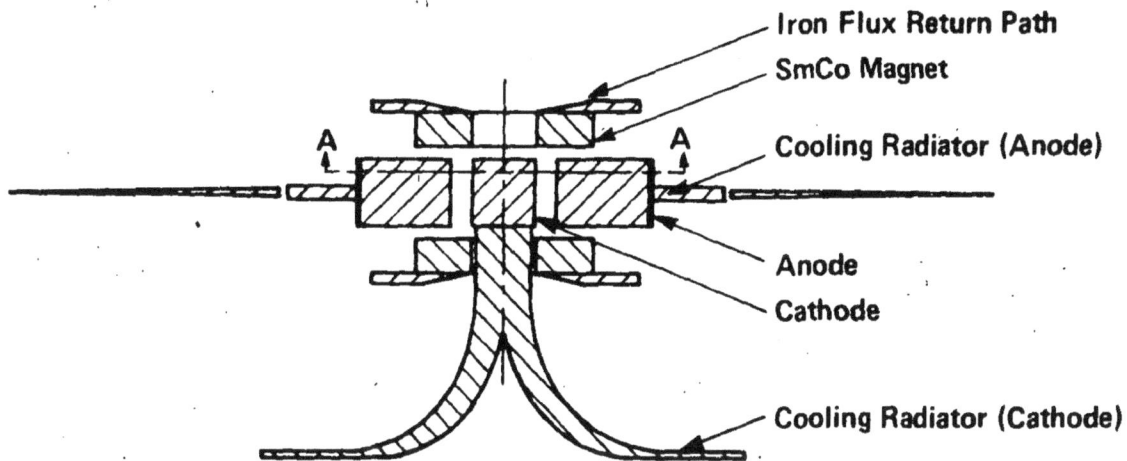

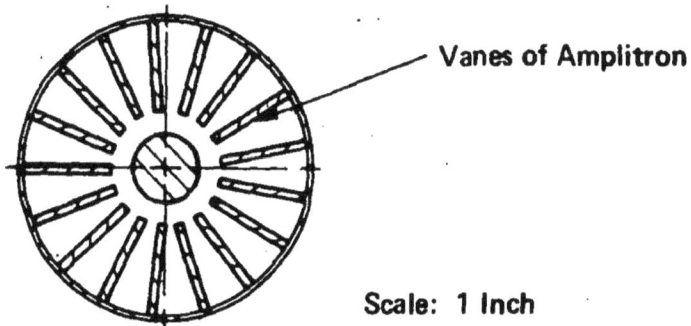

Tube Operating Parameters:
  Wavelength — 15 cm
  A. V. — 20 kV
  B/B_0 Ratio — 10
  $180°$ Circuit Mode

FIGURE 57. — SSPS AMPLITRON (Initial Design)

80

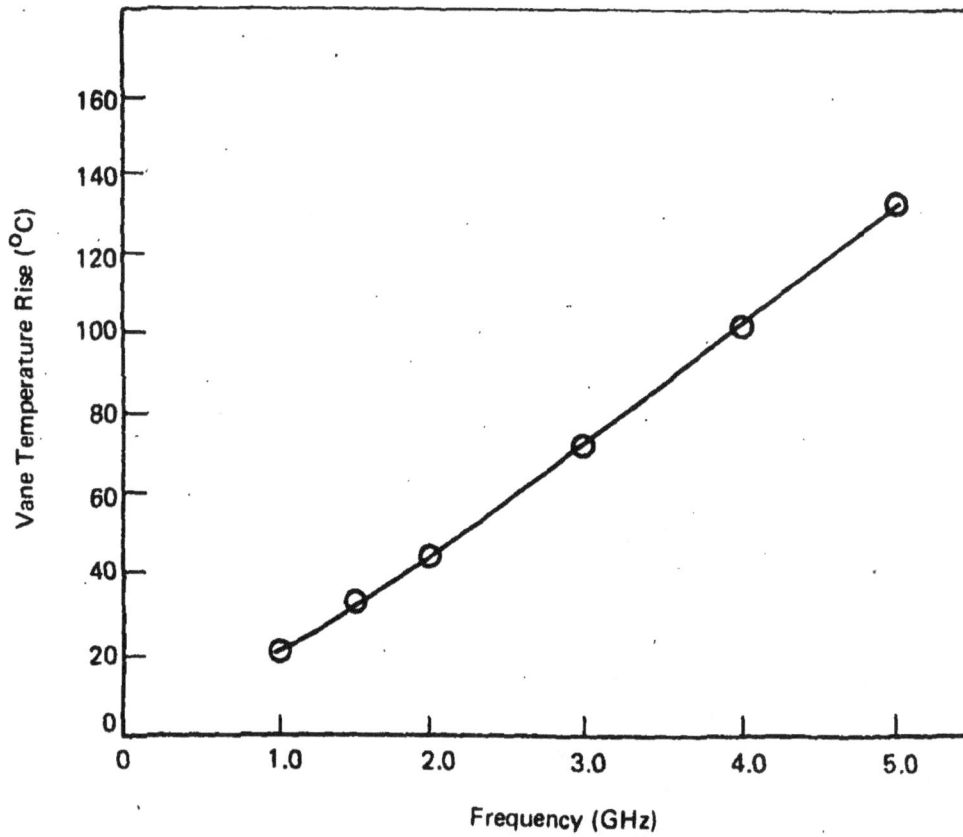

FIGURE 58. – VANE TEMPERATURE RISE VERSUS FREQUENCY
IN SSPS AMPLITRON (Initial Design)

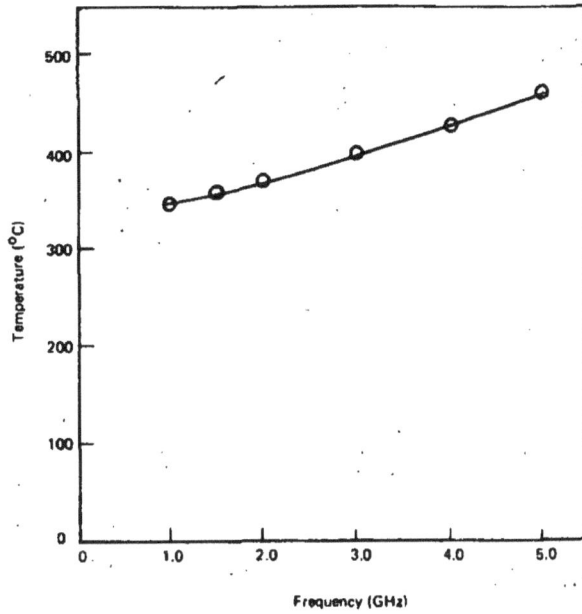

FIGURE 59. — VANE TEMPERATURE VERSUS FREQUENCY OF SSPS
AMPLITRON

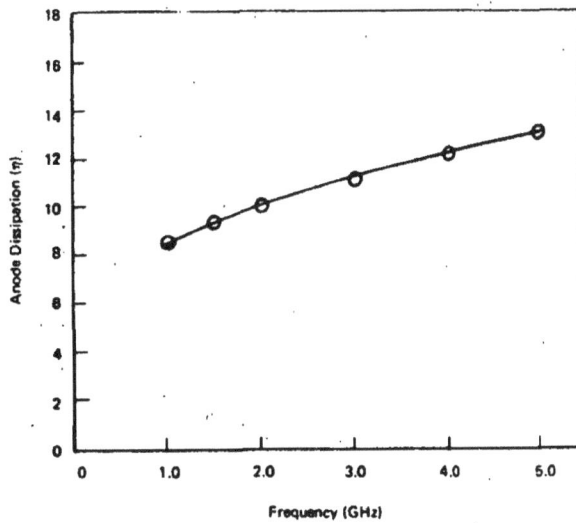

FIGURE 60. — ANODE DISSIPATION VERSUS FREQUENCY

82

The magnetic circuit of a typical crossed-field microwave tube represents a significant part of the mass and weight of the complete device. The magnetic field, B, required for the SSPS Amplitron is illustrated in Figure 61. At the higher frequencies, higher magnetic fields are required. With the development of a new magnet material, samarium-cobalt, new light-weight magnetic circuits can be developed.

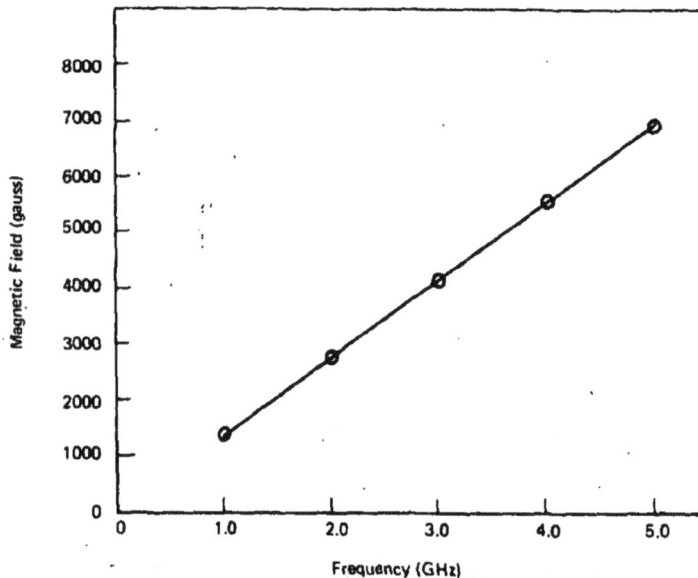

FIGURE 61. – MAGNETIC FIELD VERSUS FREQUENCY

The significant advantage realized with samarium-cobalt is illustrated by the design of an ultra light-weight magnetic circuit for an S-band 8129-type Amplitron utilizing samarium-cobalt (Sm-Co) permanent magnets (as shown in Figure 62). The Sm-Co magnet weighs 10 pounds. The present Alnico V magnetic circuit weights 89 pounds. Using the Sm-Co magnetic circuit, the weight of the S-band 8129-type CFA can be reduced from 114 pounds to 35 pounds.

A computer program was utilized to design Sm-Co magnetic circuits for the SSPS Amplitron at the main frequencies of interest, 2.0, 2.45, and 3.3 GHz. The design of the magnetic circuit is important in achieving a light-weight tube. Also, the heat flow path of the cathode will depend upon the size of the magnet, since the cathode cooling radiator will extend beyond the magnet.

The computer program was used to calculate the coordinates of all points at which the equipotential lines and flux lines crossed the grid. This information was then fed to a cathode-ray tube plotter which generated a field plot.

Radially gaussed samarium-cobalt magnets were investigated for an SSPS Amplitron which is to operate at 2.0 GHz. The equipotential surfaces and flux lines calculated in Figure 63 are for one quadrant of the magnetic circuit. The ordinate of the plot is the tube axis and the abscissa is the tube's mid-plane. This magnetic circuit develops the required 2700 gauss in the interaction region with a 6.08-oz. Sm-Co magnet. A pole piece is utilized for field shaping, and a steel flux return path is used.

ALNICO V
MAGNET

------- Radially Gaussed Sm-Co Magnet

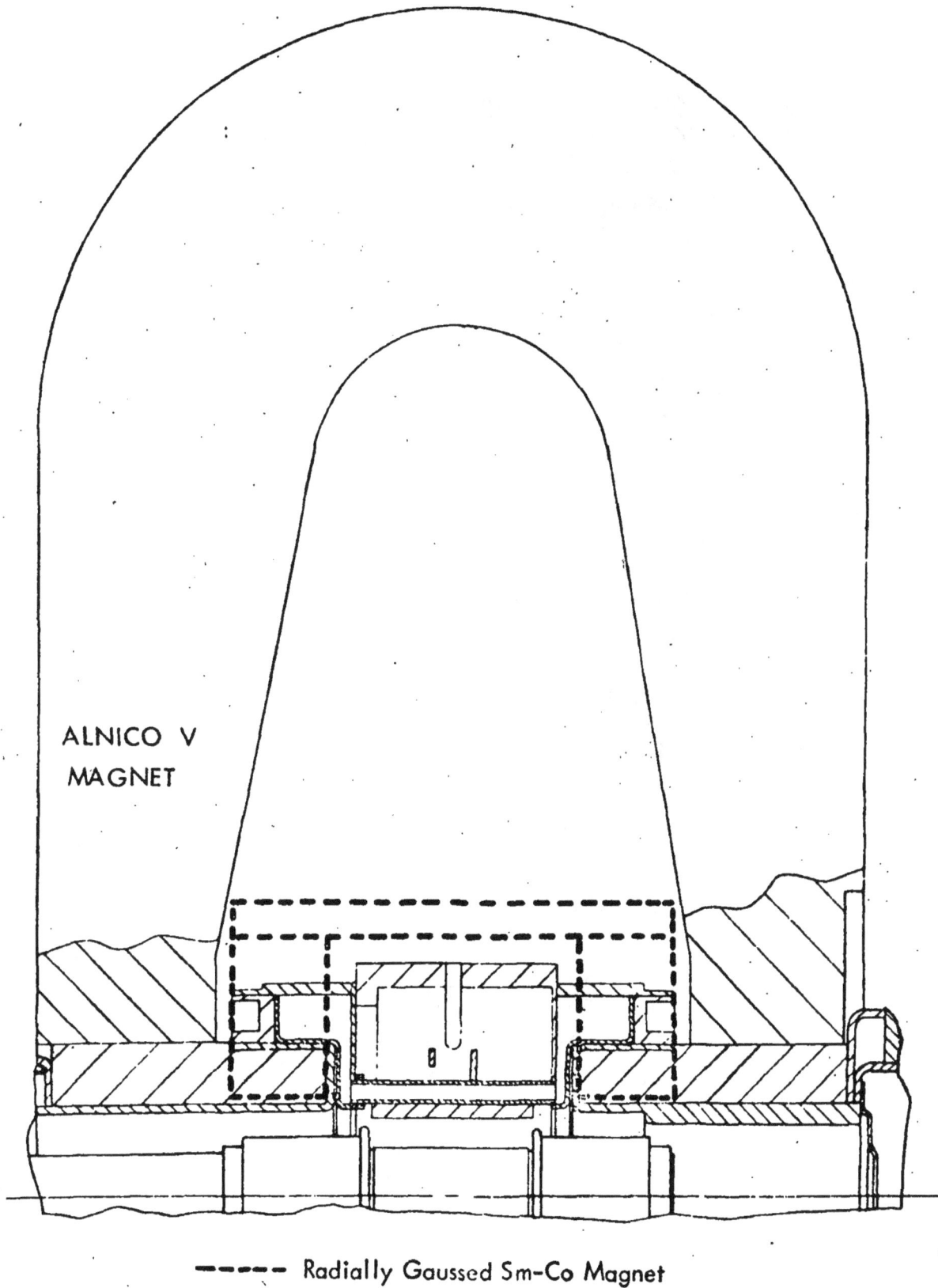

FIGURE 62.— SIZE COMPARISON BETWEEN THE RADIALLY GAUSSED Sm-Co
MAGNET AND THE ALNICO V MAGNET FOR 8129 TYPE CFA

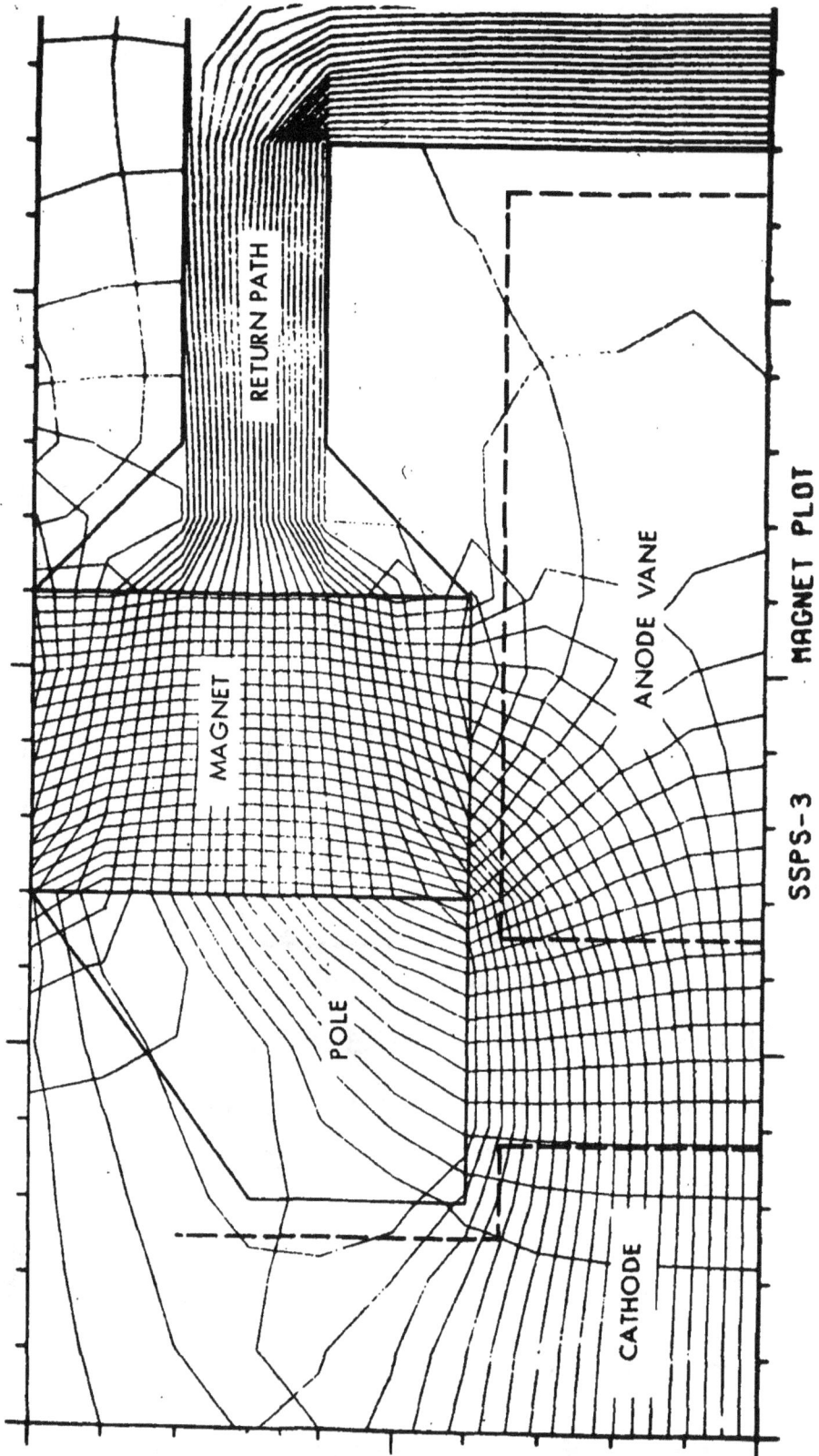

SSPS-3   MAGNET PLOT

FIGURE 63. – MAGNETIC CIRCUIT AMPLITRON AT 2.0 GHz

85

A magnetic circuit for 2450 MHz, as shown in Figure 64, was also designed. The required 3.38 kG can be obtained with this magnetic circuit. The Sm-Co magnet weighs approximately 6.25 oz with a magnetic height of 1.47 cm. The cathode back-bombardment power of 150 watts dissipated in the cathode requires an external radiator for its radiation into space. The radiator is conductively coupled to the cathode resulting in a cathode heat flow path which is approximately 0.3 cm longer than the magnet. The pyrolytic graphite radiator will be approximately 0.8 of the cathode diameter, but could be tapered larger at the top if necessary. With a thermal path of 2.4 cm, the cathode ΔT will be 132°C for the 2450-MHz Amplitron. By tapering the heat flow channel, the area could be increased more than half way, so that the ΔT could be reduced by approximately one third to 88°C.

The magnetic circuit of the SSPS Amplitron operating at 3.3 GHz is shown in Figure 65. This radially-gaussed samarium-cobalt magnetic circuit supplies the required 4.55 kG and a desirable field shape in the interaction space. The magnet is 1.7 cm high and weighs approximately 6.26 oz. The cathode ΔT is 243°C and can be reduced to approximately 161°C by tapering the heat flow channel.

The ΔT in the cathode will not be a serious problem at either 2.45 or 3.3 GHz. Although the magnetic flux requirement is greater at the higher frequency, the volume and magnetic gap at 3.3 GHz results in the weight of the Sm-Co magnet being the same as at 2.45 GHz. The magnetic circuit for the SSPS Amplitron may consist of radially-gaussed Sm-Co magnets (which weigh approximately 6.26 oz), pole pieces for field shaping, and a steel flux return path (which would add approximately 3 oz to the weight of the magnetic circuit). Additional work on the magnetic circuit could result in a reduction of the magnetic weight by approximately 25%.

Experience has shown that it is desirable to have the pole pieces for field shaping on the cathode structure to increase the operating efficiency of CW tubes. This results in a sharp knee on the V-I curve, which is necessary for CW tubes that operate at a low plate current.

A computer plot of a Sm-Co magnetic circuit where the pole piece is attached to the cathode is shown in Figure 66. This is represented by the gap between the field shaping pole piece and the Sm-Co magnet. This magnetic circuit develops the required 4550 gauss in the interaction region for an SSPS Amplitron operating at 3.3 GHz. The weight of this radially-gaussed Sm-Co magnet is 5 oz.

The study showed that the SSPS Amplitron could operate at either 2.45 or 3.30 GHz. Pertinent parameters are presented below:

	2.45 GHz	3.3 GHz	Max Temp.
Vane temperature	380°C	405°C	460°C
Cathode temperature	413°C	486°C	600°C
Sm-Co magnet weight	6.26 oz	6.26 oz	--

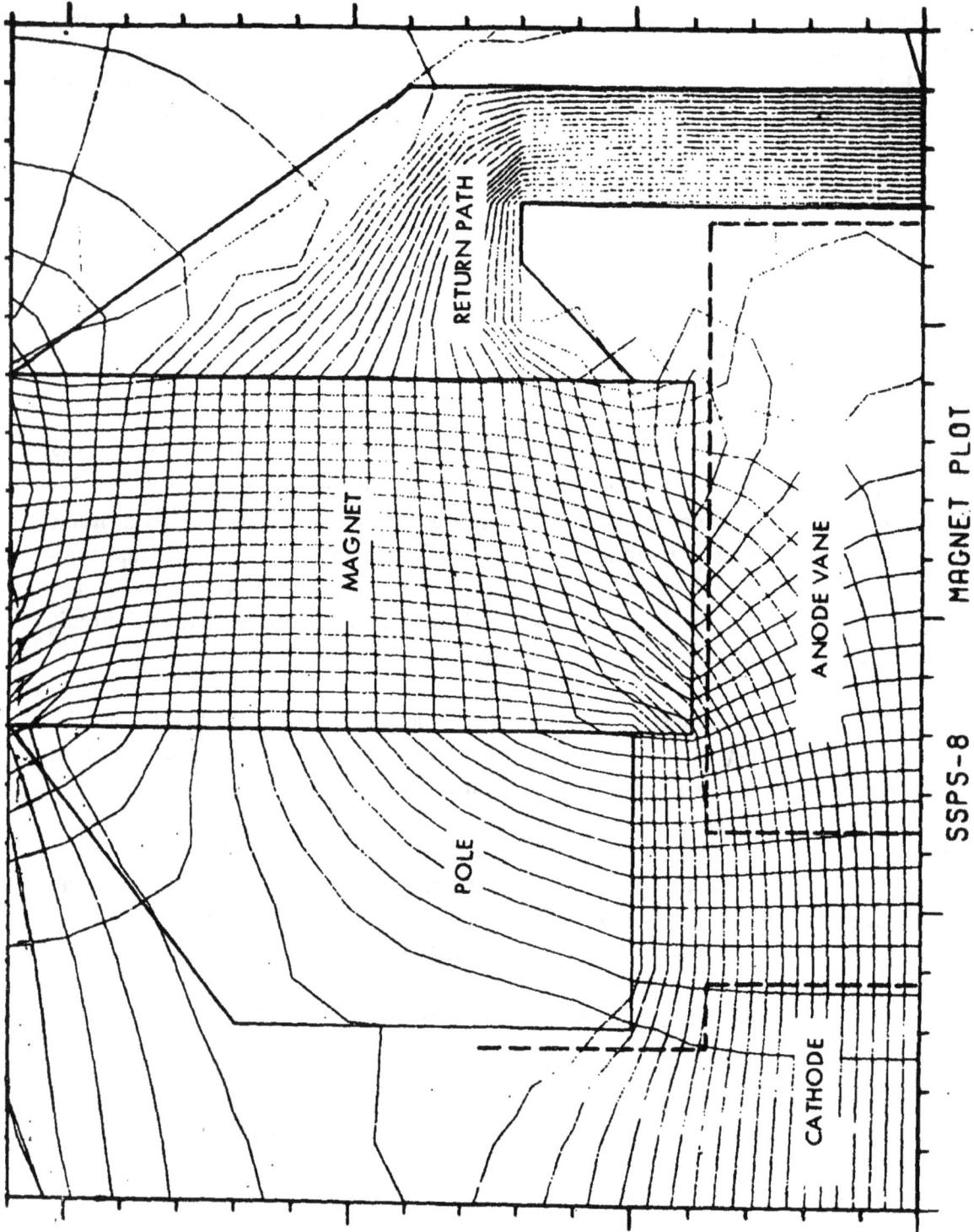

FIGURE 64. – MAGNETIC CIRCUIT OF AMPLITRON AT 2.45 GHz

SSPS-8   MAGNET PLOT

RETURN PATH

MAGNET

POLE

ANODE VANE

CATHODE

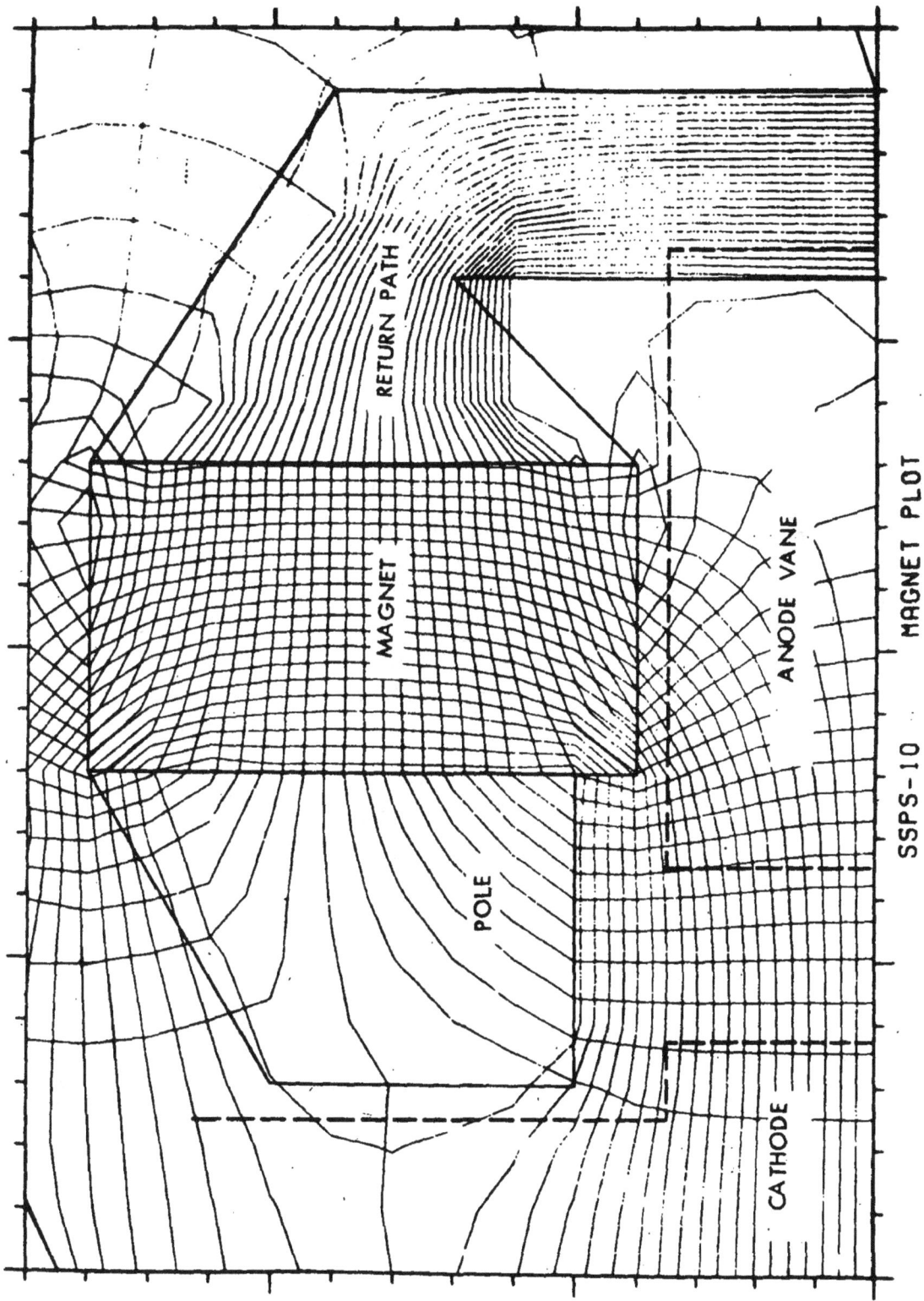

SSPS-10 MAGNET PLOT

FIGURE 65. — MAGNETIC CIRCUIT OF AMPLITRON AT 3.3 GHz

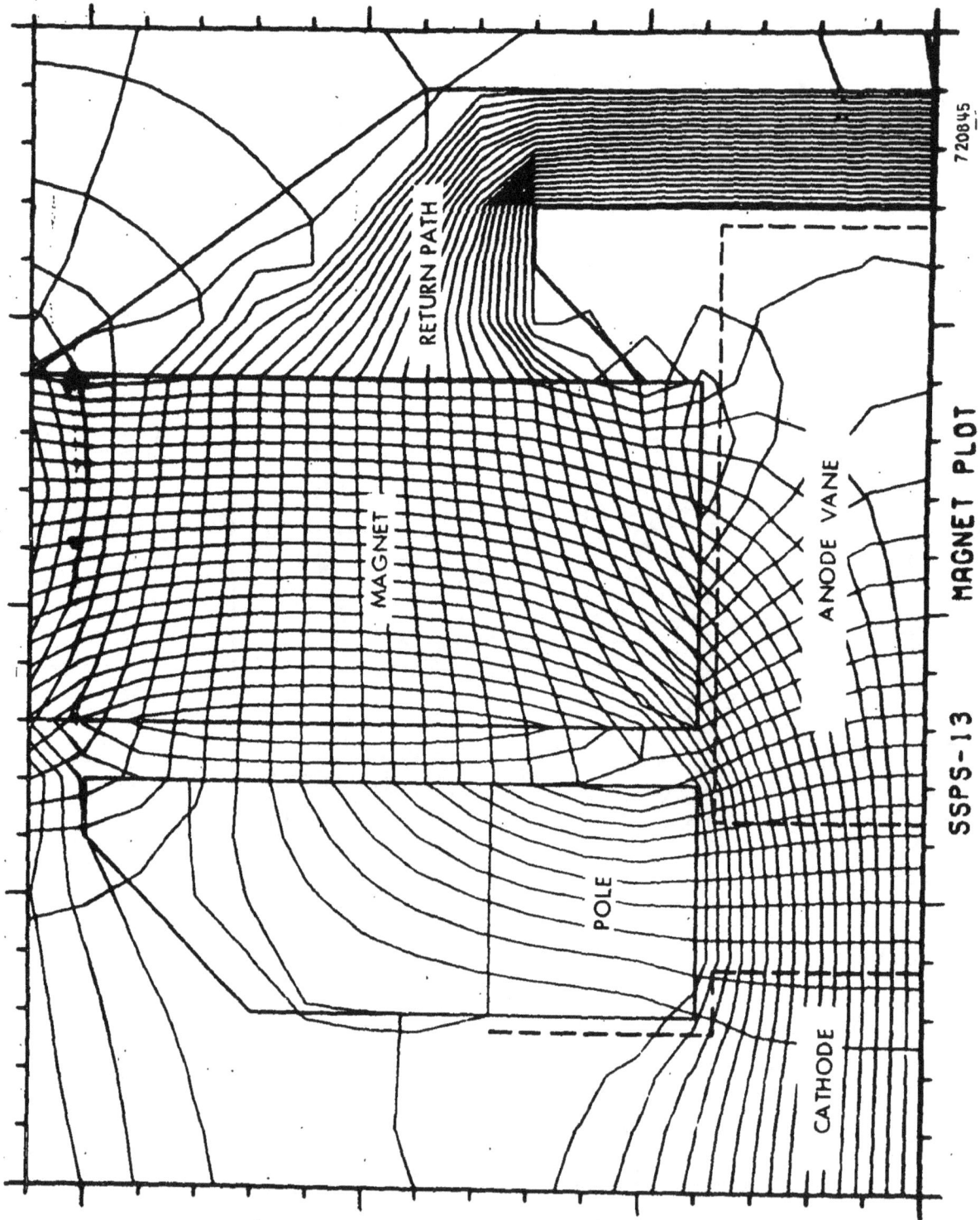

FIGURE 66. – MAGNETIC CIRCUIT OF AMPLITRON AT 3.3 GHz (Pole Piece Attached to Cathode)

89

Further study should be considered in the following areas:

- Materials — low circuit loss; higher temperature;
- Lighter weight magnetic circuits;
- Field shaping for higher efficiency;
- Space environment — vacuum and thermal; and
- Higher temperature performance of Sm-Co magnets.

The ambient temperature for the cathode pyrolytic graphite radiator can be higher than that for the anode. The cathode utilizes high-temperature metals such as platinum and molybdenum; therefore, the cathode temperature may be 600-800°C. This would put the pyrolytic graphite radiator at a temperature of about 500°C. The temperature of the anode would be limited by the silver solder to approximately 450°C. The samarium-cobalt magnets would probably limit the temperature at which passive cooling may take place. The upper operating temperature on the Sm-Co magnet material is in the order of 350°C because of the loss of gauss with temperature.

RF Spectrum Considerations for the SSPS Amplitron. — The cross-field amplifier (Amplitron) can produce the following microwave energy emissions: (a) the main RF signal which carries the power; (b) harmonic energy which relates to the main signal and is produced by slight distortion in the generating process; (c) energy that occurs during the turn-on and shut-down sequence; (d) random background energy that is intrinsic to the electron interaction mechanism in the tube; and (e) spurious signals, unrelated to the others, which may arise as a result of the specific design.

The main signal energy is the primary energy produced by the tube for the system function. This signal is locked onto the RF drive signal, and, under DC or CW conditions, its spectrum would be characterized by a single frequency, phase-locked or controlled by the drive signal. Any movement, variability, or modulation of this signal would be related entirely to the drive signal source quality. Any additional variability or spectrum components produced by the amplifier would arise only from instability of the dc power source.

With regard to harmonic energy, we expect that all harmonics produced by the tube would be at least 45 dB below the main signal amplitude. Harmonic energy occurs because of the non-linearity of the high efficiency saturated interaction process within the tube. Spectrum components are exact multiples of the main signal frequency, and their intensity is influenced in part by the properties of the various elements of microwave plumbing through which the signal passes. The most complete data on harmonics applicable to the crossed-field amplifier (CFA) were taken on a radar system built by ITT-Gilfillan. The system consists of a chain of two Amplitrons and a traveling-wave tube driver (Figure 67).

Typical data presented (Figure 68) show the strength of the second and third harmonics under several operating conditions. The harmonics measured at the output with the final and driver-stage Amplitrons off are lower, partly because the passband of the circulator and the Amplitrons perform a filter function as the signal passes through. When the entire chain is on, least filtering occurs on the final stage and the harmonics are strongest, although as low as 45 dB down. These data were

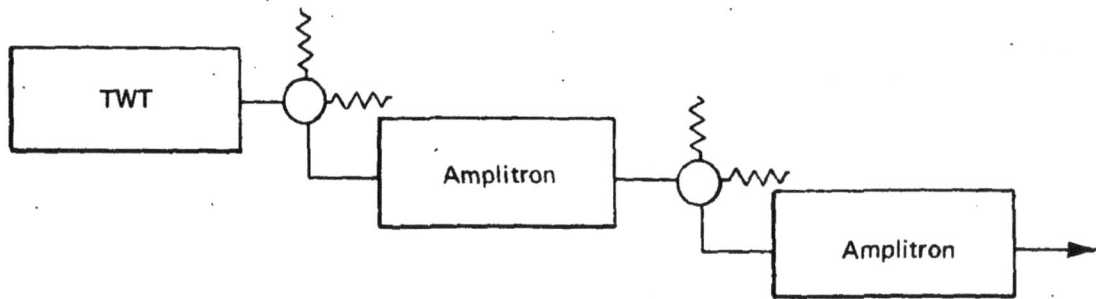

FIGURE 67. — SCHEMATIC OF ITT GILFILLAN-BUILT RADAR SYSTEM

obtained in a microwave configuration having no special provision for filtering or diminishing the harmonics. These would appear to be typical data and can be used as a base for assessing the probable performance of the proposed SSPS tube.

The third kind of energy – energy occurring during turn-on and shut-down – is produced at such times because the voltage applied to the tube traverses values which are coincident with other electronic interacting modes of the tube. Since it is not physically possible to swing the voltage from its running value to zero in zero time, there will be a finite amount of time spent at a possible oscillating condition. This time is determined entirely by the transient properties of the dc power source. The frequency of the oscillation will be approximately 15% below the running frequency. In pulsed radar tubes, the voltage passes rapidly through the oscillating value on every pulse and a short burst of energy results. When the rate of rise and fall of the voltage pulse is fast, no energy is produced. However, in practice, the 15% frequency separation makes it feasible to filter out whatever energy does occur. Under dc operating conditions, the voltage, of course, always remains at the running value and no signal of this type is produced in normal steady operation.

The last type of energy can be categorized as discrete spurious energy related to microwave resonances in the tube and its environment. The microwave properties of a typical CFA are shown in Figure 69, which indicates a well-matched, operating passband in the mid-region of frequency and mismatches or resonances in the outer regions of the band. Sharp resonances tend to peak up background noise energy and produce concentrations of it at certain discrete frequencies.

A spectrum analyzer presentation of this noise (Figure 70) was taken on a pulsed tube having a platinum cathode similar to the one under consideration for the proposed tube. Precise measurements of this noise have been made recently on a number of pulsed tubes with the data taken between the spectral lines of the signal and mathematically converted to an equivalent CW value.

Table 15 presents CFA spectrum noise measurements. The best values obtained are about 55 dB/MHz below the main signal and were taken near the limits of the instrumentation. There is some indication that the intrinsic limit in the CFA may be better than presently observed, but this has not yet been confirmed. Very sketchy data on CW operation are available. These consist of one test on a low-power amplifier and another on a high-power oscillator. The oscillator data are consistent with the pulsed measurements, while the low-level amplifier data are about 30 dB better. Further data are needed.

91

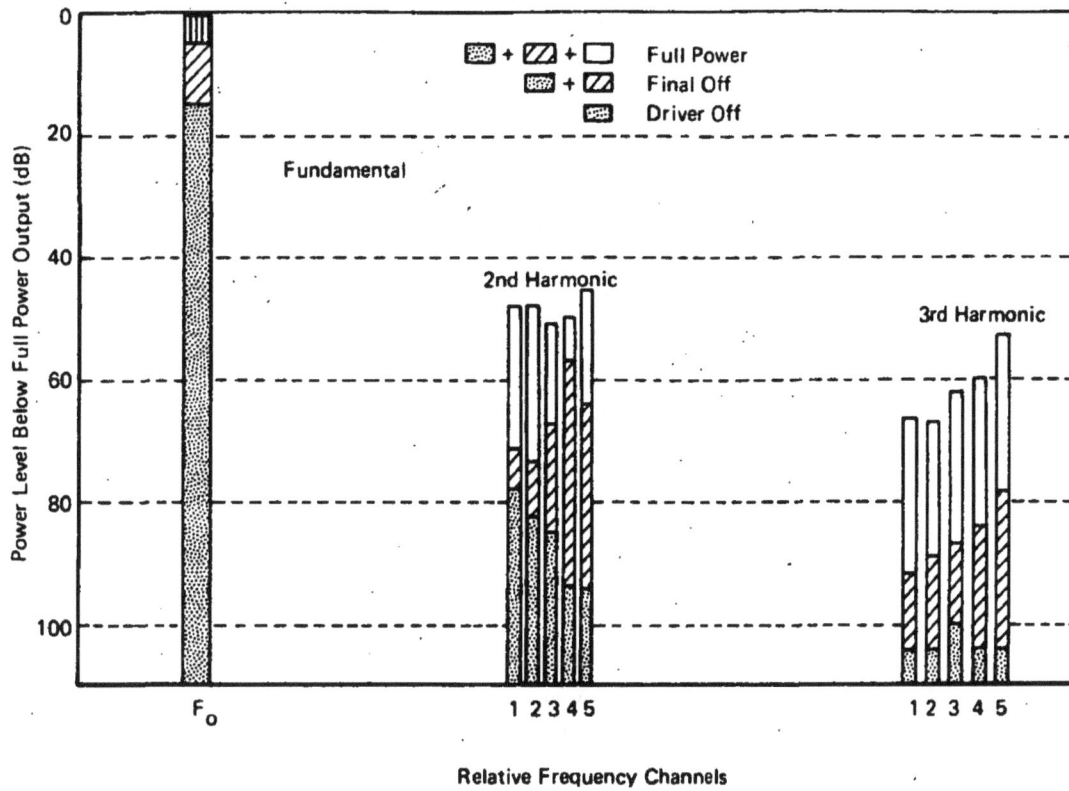

FIGURE 68. – FIXED FREQUENCY HARMONIC OUTPUT - NORMAL WAVEGUIDE SYSTEM

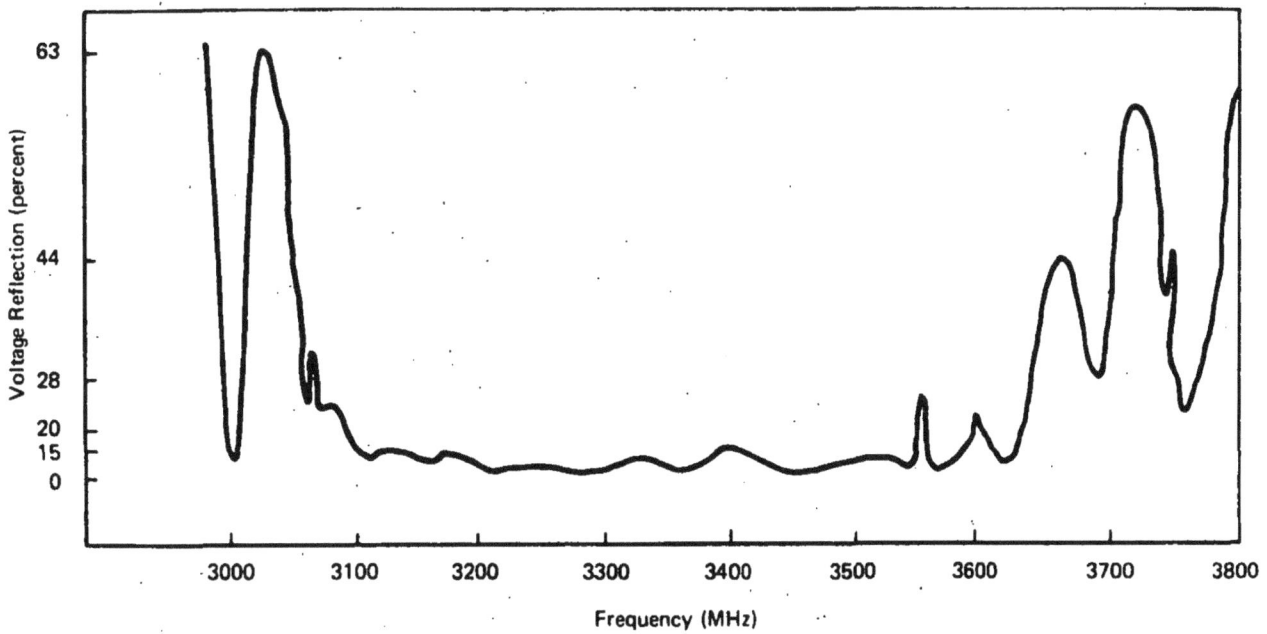

FIGURE 69. – QKS 1646 NO. 2 IMPEDANCE MATCH

92

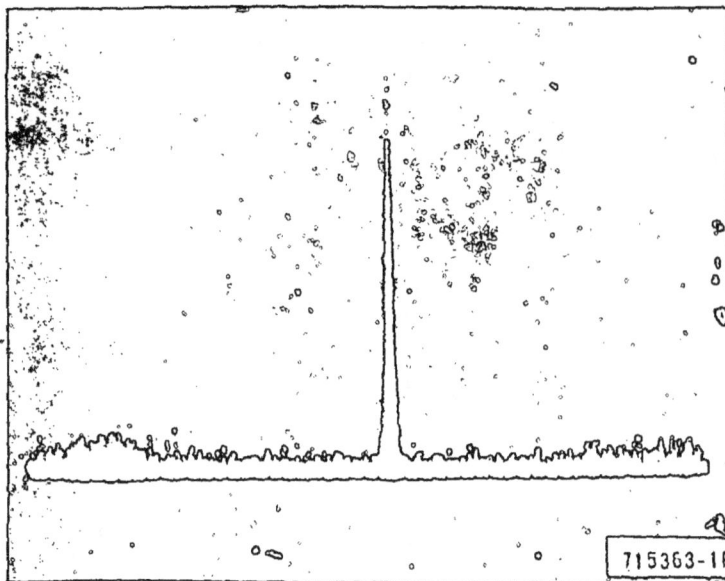

715363-1P

FIGURE 70   SPECTRUM ANALYZER PRESENTATION
OF TURN-ON/SHUT-OFF TYPE ENERGY

TABLE 15

CFA INTRA-SPECTRUM NOISE MEASUREMENTS

		SIB/NIB* (dB)	S/N** (dB/MHz)
L-Band	(Sanders)	71.8	-40.2
FW-CFA QKS1319		67.0	-42.4
L-Band	(Raytheon)	75.4	52.4
FW-CFA QKS1319			
C-Band	(Raytheon)	69.0	55.0
QR1606	(Raytheon)	85.0 to -87.0	-55.0 to -58.0

*Measured noise power relative to spectral line power
**"Equivalent CW" noise power density

93

Putting all of the microwave emissions together into a single composite, the spectrum envelope near the main signal would be as shown in Figure 71. The three lobes in the center indicate the range of frequency over which the allocation for the main signal might be, for example, a 100-MHz region centered on the 2450-MHz industrial heating band. The three lobes centered on -400 MHz are the turn-on signals associated with each of the three running frequencies. These signals are non-existent during steady operational running. The solid envelope line is the approximate noise level for the random background noise as a function of frequency derived from measurements on a pulsed cold platinum cathode type of CFA. The solid-line envelope represents the current state of affairs in existing CFA's which have had no specific effort directed in their design or application to minimize or improve the RFI emission.

The microwave properties of a typical CFA indicate a well-matched, operating passband in the mid-region of frequency and mismatches or resonances in the outer regions of the band. Sharp resonances tend to peak up background noise energy and produce concentrations of it at certain discrete frequencies. Most of them are the result of the metallic confines of the tube necessitated by the vacuum envelope in the natural atmosphere of the Earth. In a space application where the environment is under vacuum, the necessity for such an enclosure is removed and much of this resonance phenomenon will not exist. In any case, resonances can be considered a natural and routine problem that is encountered in every development for which relatively straightforward design steps are available.

Looking towards what might be achievable with specific effort directed at optimization for the space requirement, certain beneficial trade-offs are possible which have not been required nor implemented for radar systems usage. Most of the presently available tubes have been built for usage in broadband radar. The design trade-off has been geared to other requirements, notably gain and bandwidth, rather than toward the RFI problem. This means that trade-offs can be made that have not been implemented before because they were not necessary, or they were not possible for other reasons.

One of the available trade-offs for this requirement results from the narrow-band or single-frequency operation. This means that tube performance need not be optimized over the frequency band. The optimization can be made over a narrow frequency range. With a narrower passband, the frequencies on either side of the main signal would be filtered more than they are in current designs, which would reduce, appreciably, the noise level away from the main signal. The outlook, therefore, would be to get improvements in the spectrum contour by tailoring the tube to this requirement. With additional filtering, it could be improved even more. The general tube-design concept would be to use all of the latitude available in the narrow-band space environment application, such as the use of unenclosed structures to eliminate resonant modes, no input or output windows with their associated resonances, and low gain which has the further effect of suppressing other signals. The trade-off would focus on the RFI question rather than upon the standard requirements in Earth-bound applications.

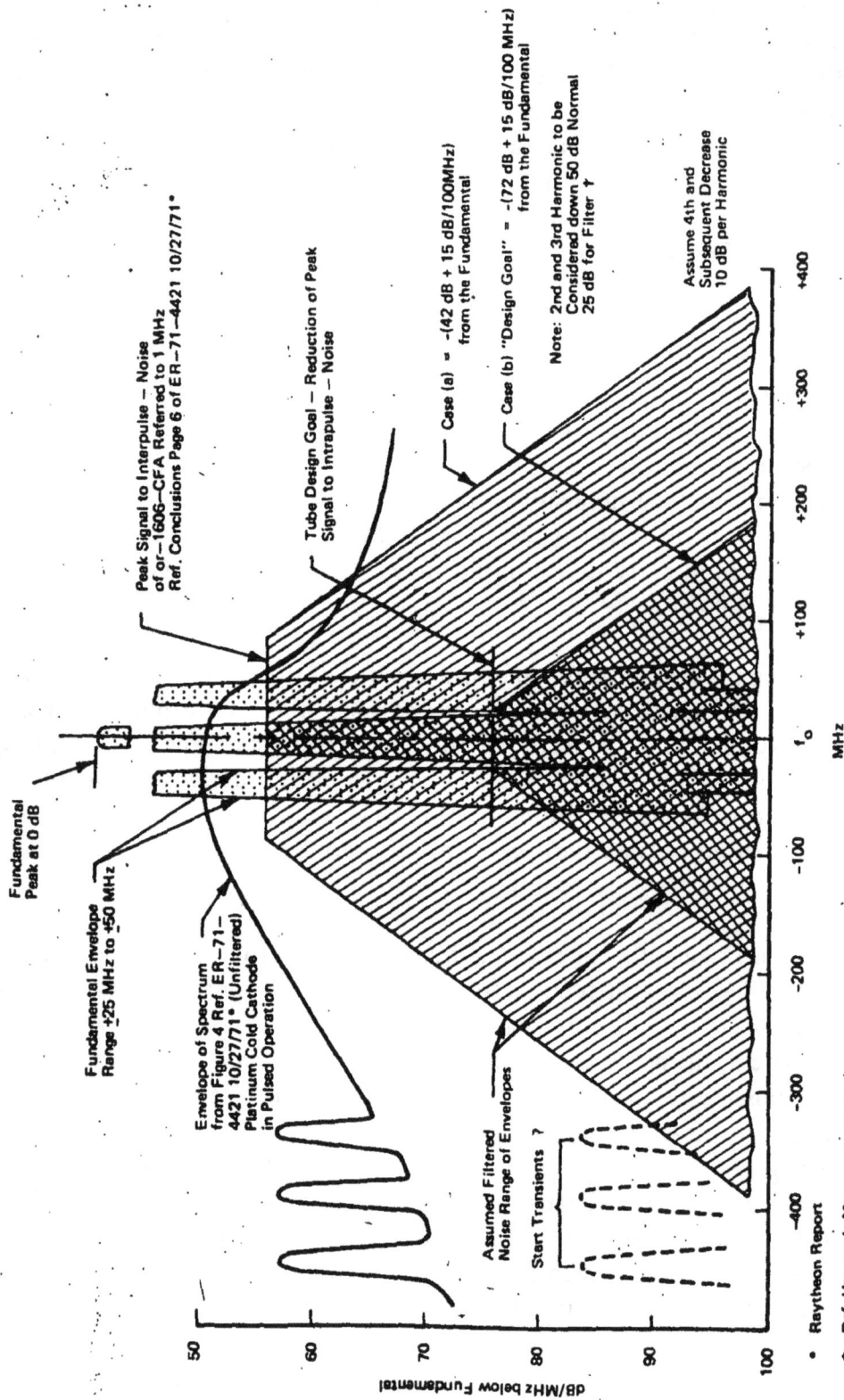

FIGURE 71. — RF SPECTRUM ASSUMPTIONS FOR SSPS INVESTIGATIONS CLOSE TO MAIN BEAM

95

With respect to additional filtering, which would be needed for the suppression of harmonics, there is a waveguide type known as the "waffle iron" which appears to be most adaptable to this requirement. The design literature indicates that a 50 to 70 dB attenuation level over a 10 to 1 bandwidth is achievable.

There are areas of further study that are suggested to resolve or further define the problem. These include: (1) narrow-band operation in an unusual design requirement, which warrants some further study to confirm the impact upon noise; (2) the selection of the optimum gain which could range from 3 to 10 dB for each stage should be studied to identify the best overall value for the RFI requirement; (3) noise measurements on CW tubes should be made to confirm the interpretation of present pulsed data converted to CW (there is an expectation that such data will, in fact, be better than that obtained on pulsed operation); and (4) filter capability in relation to size, weight, and attenuation tradeoff.

Effects of SSPS RFI on Other Users

Transmitting Antenna and Nature of the Transmitted Beam. – The geometry of the transmitting and receiving antenna system is shown in Figures 72 and 73, and Table 16 presents the electrical characteristics of the transmitting antenna. A preliminary investigation of the amplitude distribution for minimal Earth spot which contains 90% of the total energy shows the distribution to be $[1 - r^2]^{1/2}$ where r is the radial point on the transmitting antenna. The power distribution pattern for the transmitting antenna is shown in Figure 74, and the resulting pattern for the receiving antenna is shown in Figure 75. There are other distributions which should be investigated later in the program that can be used to reduce the size of the Earth spot even further; however, these are more complicated and, at best, will only cause a reduction of a few percent. (The receiving antenna is also discussed in this section.) The distributions of the form $[1 - r^2]^n$ are conveniently analytical and are perfectly adequate at this stage of the analysis. System data are summarized in Table 17 for n = 0, 1/2, 1, and 2.

TABLE 16

ELECTRICAL CHARACTERISTICS OF THE TRANSMITTING ANTENNA

Frequency of operation	= 3300 MHz (λ = 0.091 meter)
Total power into transmitting antenna	= 6.4×10^9 watts (97 dBw)
Gain of transmitting antenna	= 8.7×10^8
Number of Amplitrons in transmitting antenna	= 8×10^5
Power distribution in transmitting antenna	= $(1 - r^2)^{1/2}$

96

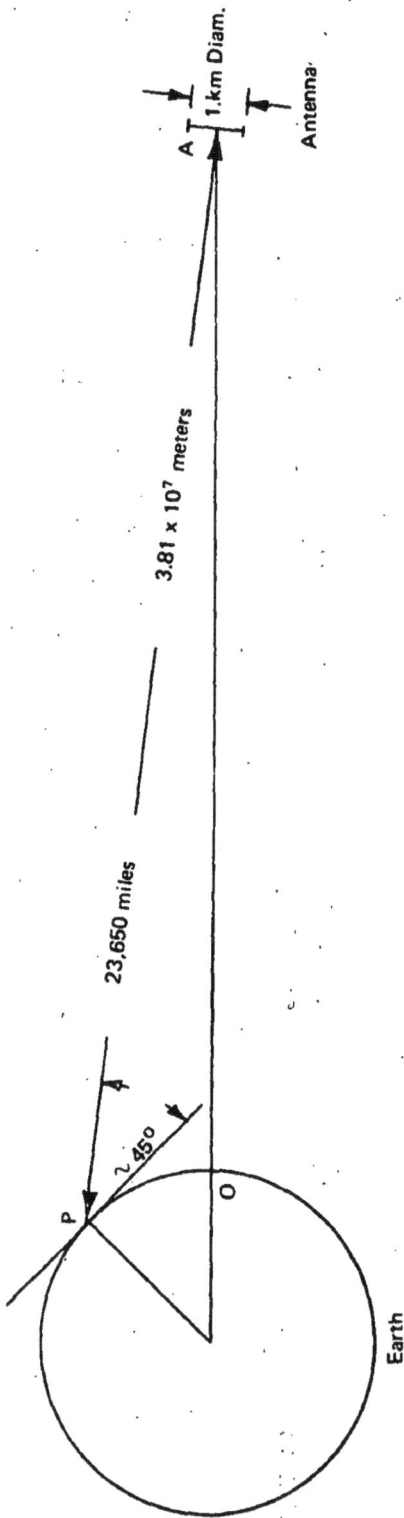

FIGURE 72. – ANTENNA-EARTH SURFACE GEOMETRY
(U.S. location)

1.km Diam.

A

Antenna

3.81×10^7 meters

23,650 miles

$\sim 45°$

P

O

Earth

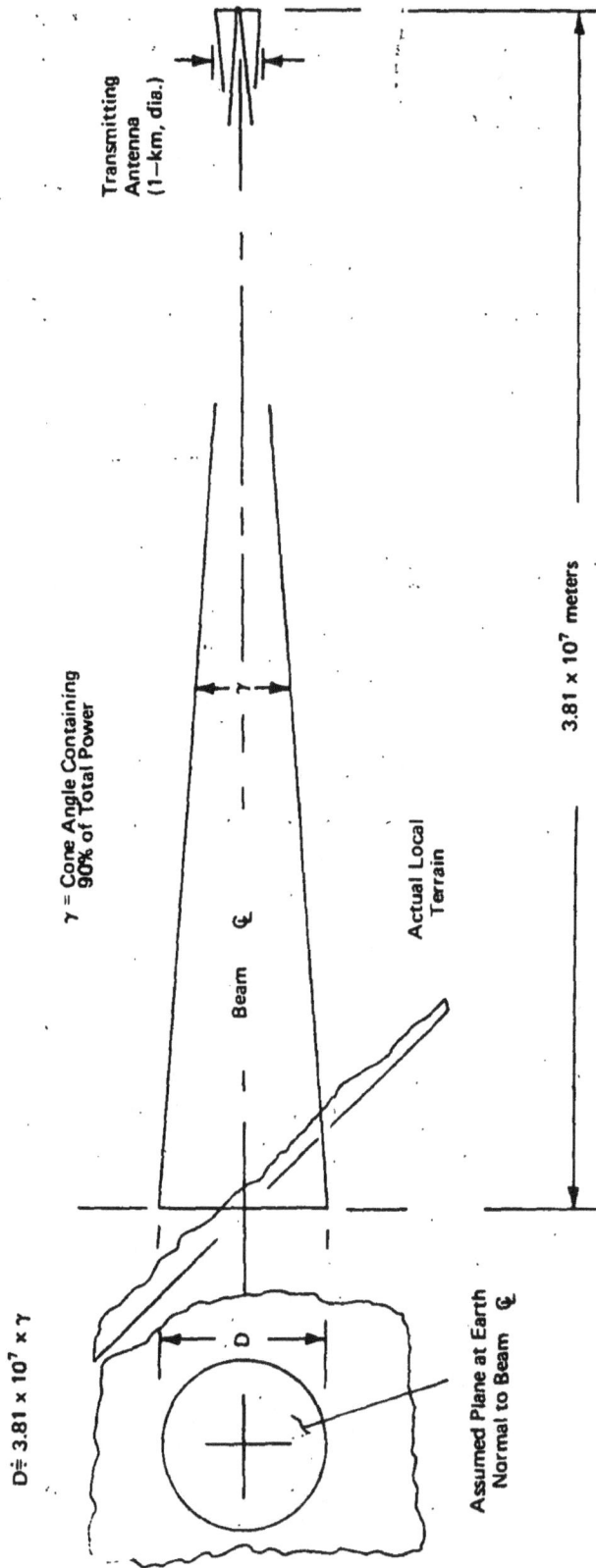

FIGURE 73. RECTIFYING ANTENNA REFERENCE PLANE NORMAL TO BEAM ₵

Transmitting
Antenna
(1–km, dia.)

3.81×10^7 meters

γ = Cone Angle Containing
90% of Total Power

Beam ₵

Actual Local
Terrain

$D \doteq 3.81 \times 10^7 \times \gamma$

D

Assumed Plane at Earth
Normal to Beam ₵

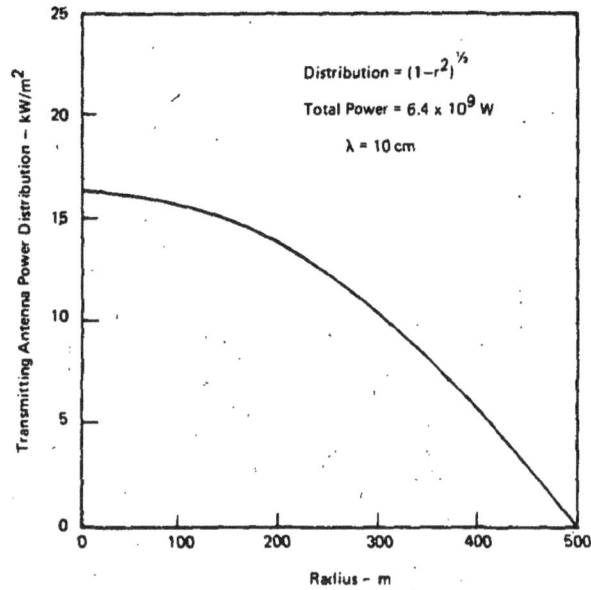

FIGURE 74. — TRANSMITTING ANTENNA EXPONENT = ½ POWER DISTRIBUTION

The figure shows Transmitting Antenna Power Distribution – kW/m² versus Radius – m, with:

Distribution = $(1-r^2)^{1/2}$

Total Power = 6.4×10^9 W

$\lambda = 10$ cm

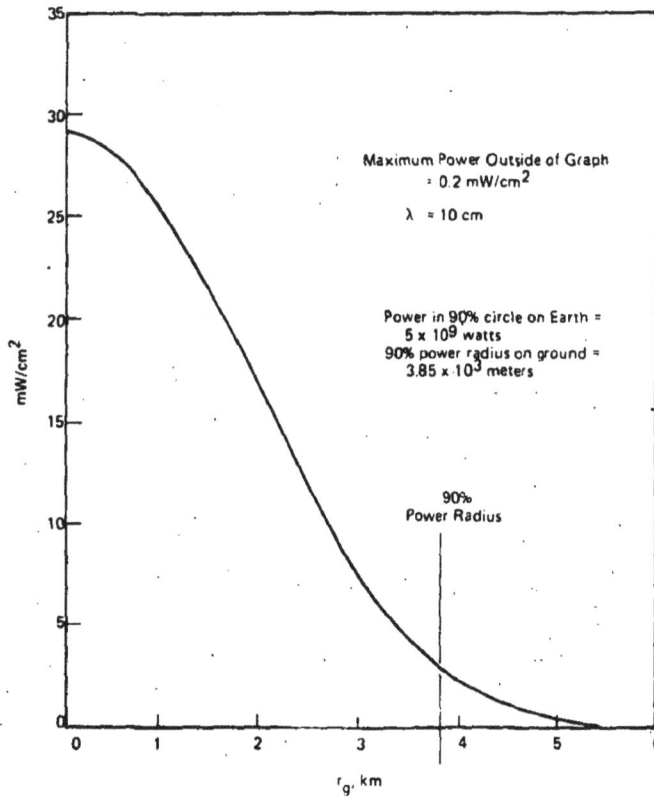

FIGURE 75. — RECTIFYING ANTENNA POWER DISTRIBUTION FOR EXPONENT = ½ DISTRIBUTION AT TRANSMITTING ANTENNA

The figure shows mW/cm² versus r_g, km, with:

Maximum Power Outside of Graph = 0.2 mW/cm²

$\lambda = 10$ cm

Power in 90% circle on Earth = 5×10^9 watts

90% power radius on ground = 3.85×10^3 meters

90% Power Radius

98

TABLE 17
SUMMARY OF POWER DENSITIES DISTRIBUTIONS AND RADII

	PARAMETER	DISTRIBUTION EXPONENT*				UNITS
		0	½	1	2	
2nd Side Lobe	Power Density	.138	.035	.011	.0018	mW/cm²
	Radius	10.	11.1	11.8	≈ 15	km
1st Side Lobe	Power Density	.60	.20	.083	.016	mW/cm²
	Radius	6.1	6.9	7.6	9.25	km
Main Lobe	Power Density on Axis	33.6	29.3	25.2	18.8	mW/cm²
	Power Density at 90% Power Rad.	.25	2.7	2.7	1.6	mW/cm²
	Power/Rectifier at 90% Power Rad.	12.9	140.	140.	83.	mW/Rectifier
	Radius to 90% Power	7.35	3.85	4.06	5.0	km
	Curve Shape	(curve)	(curve)	(curve)	(curve)	−33.6 mW/cm² −29.3 mW/cm² −25.2 mW/cm² −18.8 mW/cm² ** < .01 mW/cm² for r > 15 km
CENTER		⊢5km⊣	⊢5km⊣	⊢5km⊣	⊢5km⊣	
Power Density		8.15	16.3	24.45	40.75	kW/m²
Curve Shape		(curve)	(curve)	(curve)	(curve)	−40.75 kW/m² −24.45 kW/m² −16.3 kW/m² − 8.15 kW/m²
		⊢500m⊣	⊢500m⊣	⊢500m⊣	⊢500m⊣	

RECTIFYING ANTENNA AT EARTH

TRANSMITTING ANTENNA λ = 10 cm, D_t = 1000 m, Total Power = 6.4×10^9 watts

*Distribution ≈ $(1 - r^2)$

P ≈ exponent in chart

r ≈ radius from center of transmitting antenna

**For P = ½, 1 and 2; power flux density at and beyond 15 km is estimated to be less than 0.01 mW/cm² and decreasing with distance.

Notes:

Effect of Frequency Selections:
For changes in frequency (keeping the transmitting antenna size, total power, and distribution constant), the following would apply:

Lengths at Earth are multiplied by f_o/f

Power densities at Earth are multiplied by $(f/f_o)^2$ where f_o = 3 GHz

Effect of Transmitting Antenna Diameter Modifications:
For changes in transmitting antenna diameter (keeping the total power and distribution as constant), the following would apply:

Lengths at Earth are multiplied by D/D_o

Power densities at Earth are multiplied by $(D_o/D)^2$ where D_o = 1000 meters.

Figure 72 shows that the transmitted beam hits the Earth's surface at an angle of about 45 deg. This results in an elliptical spot which has a major to minor axis ratio of about 1.4 to 1. To simplify the analysis, the spot was considered to be circular, as if it were projected on a plane at the Earth normal to the beam centerline, as shown in Figure 73.

Receiving Antenna. – The receiving antenna consists of a large array of dipole-reflector elements placed within the 90% power radius (3.85 km). The gain of each element is about 10 and has an effective area of about 50 cm.^{2}

$$A = \frac{\lambda^2}{4\pi} \; g = \frac{830}{4\pi} \approx 66 \text{ cm}^2$$

where $\lambda = 9.1$ cm, g = 10.

Spacing these elements about 0.6λ apart will overlap the effective area enough so that very little power will go through. This gives a density of about 300 elements per square meter or a total of 1.38×10^{10} elements. Since the RF energy is converted directly to dc, there is no problem of phasing the elements. Each element at the center of the array will contribute about 1.9 watts and tapers off to approximately 0.10 watt at the 90% power radius.

A summary of the properties of the antenna system for different distributions on the transmitting antenna $[(1 - r^2)^n, n = 0, 1/2, 1$ and $2]$ is presented in Table 17. (Note that the table data are for $\lambda = 10$ cm or $f = 3,000$ MHz, not 3,300 MHz. Effects of frequency selection and antenna diameter are shown in Table 17.)

Figure 76 shows the noise temperature that an antenna sees as a function of frequency. We observe that at 3,300 MHz this temperature is about 2 to 3°K for the antenna in the zenith position (antenna with a narrow beam). This indicates this area potentially to be about the worst part of the spectrum as far as interference is concerned. If all other noise is eliminated, we are working against a noise background of only -158 dBw. Noise power = KTB = 1.71×10^{-16} watts = -158 dBw where

K = Boltzmann's Constant = 1.37×10^{-23} joules/°K
T = temperature in °K = 2.5°
B = bandwidth in Hertz = 5×10^6 Hz

In radio astronomy, the limit normally identified is approximately -200 dBw.

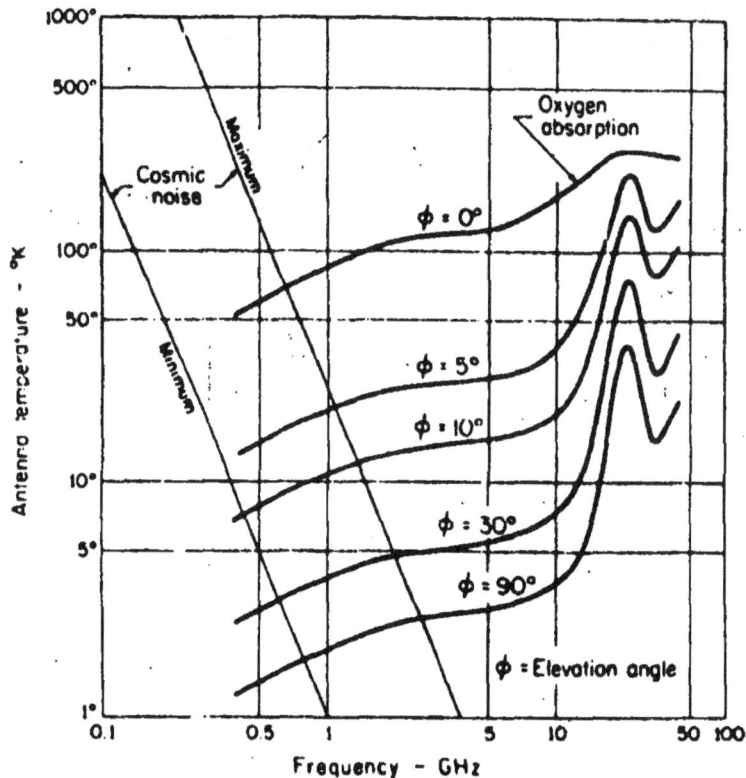

FIGURE 76. — NOISE TEMPERATURE PROFILE

Noise from the SSPS. — Diagrammatically the SSPS can be considered as a three-part system consisting of (a) a generator with output power P, (b) a transmitting antenna with gain G, and (c) a filtering system placed between the antenna and the generator.*

a. The Generator

The generator consists of a large number of Amplitrons (8×10^5), each feeding about $1\ m^2$ of the transmitting antenna surface. The noise emanating from this $1\ m^2$ is coherent, since it originates in the same Amplitron. However, it is incoherent with the noise from other Amplitrons. The total power delivered to the antenna at the design frequency (3300 MHz) is essentially 6.4×10^9 watts (+98 dBw).

The present value of the noise power is 55 dB below this level, or +43 dBw. This value is inherent in the tube design and is for a 1-MHz band, and will be included in Case (a) calculations. Case (a) is represented as presently achievable technology, and in the future it is expected that the Amplitrons can be designed so that the peak value of the noise power will be at 85 dB below the fundamental, so that the level of the noise power will be at +13 dBw. This future expectation will be included in Case (b) calculations.

---

*This is depicted in Figure 80.

b. The Transmitting Antenna

The transmitting antenna has a diameter of 1000 meters. Each square meter of surface is fed by one Amplitron. The noise radiated from this area can be considered as coherent, and at the design frequency the area has a gain of

$$G(dB) = 10 \log \frac{K4\pi A}{\lambda^2}$$

where

$A = 1 \text{ m}^2$
$\lambda = 0.091 \text{ m, and}$
$K = \text{gain constant.}$

K has a value which varies with frequency and has a peak value of about 0.83 at the design frequency and is lower the farther away one goes from 3300 MHz. This is due to many factors, including a decrease in efficiency of slot coupling, an increase in standing wave ratio, and other factors which are inherent in the antenna design. A typical curve of G versus frequency is shown in Figure 77. This curve can be approximated by the expression

$$G(dB) = 31 - 1 \text{ dB/100 MHz from the fundamental (1)}$$

c. The Filtering System

As in the case of the Amplitron, the filtering system also consists of two separate cases: Case (a), which is realizable at present, and Case (b), which is designated as the design goal or that which is achievable in the future as the result of further study.

Case (a) is divided into two parts, a2 and a4, which designate a two-section and a four-section filter, respectively. Each filter section has a half 3 dB width of 100 MHz from the fundamental. It follows the law of -6 dB per octave (of 100 MHz) per section.* Figure 78 shows a straight line curve which is a good representation of the filter performance. Figure 78 is really the resultant noise power/MHz input to the transmitting antenna, considering both the Amplitron design and the external filter design.

---

*A representative curve of this filter is shown in Figure 5, page 8-5, reference data for Radio Engineers, 5th edition, ITT.

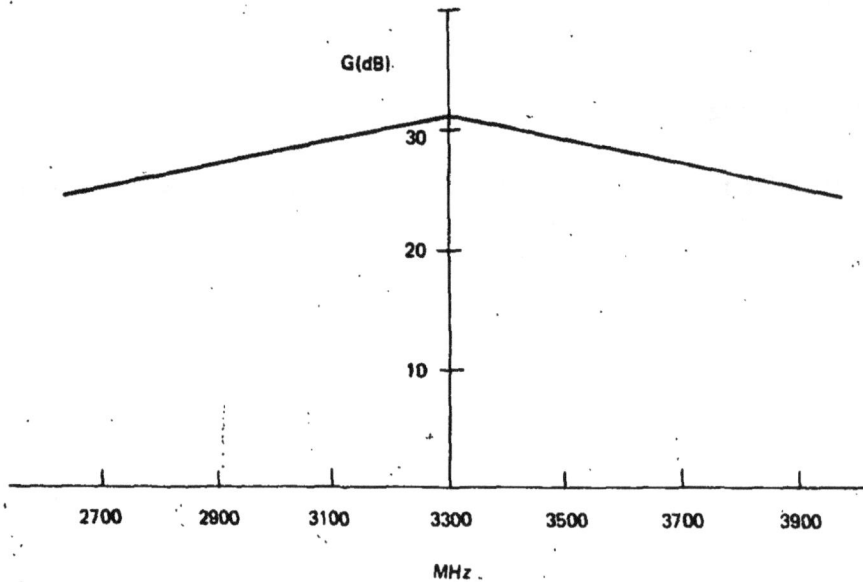

FIGURE 77. — GAIN OF SSPS ANTENNA ASSOCIATED WITH EACH AMPLITRON

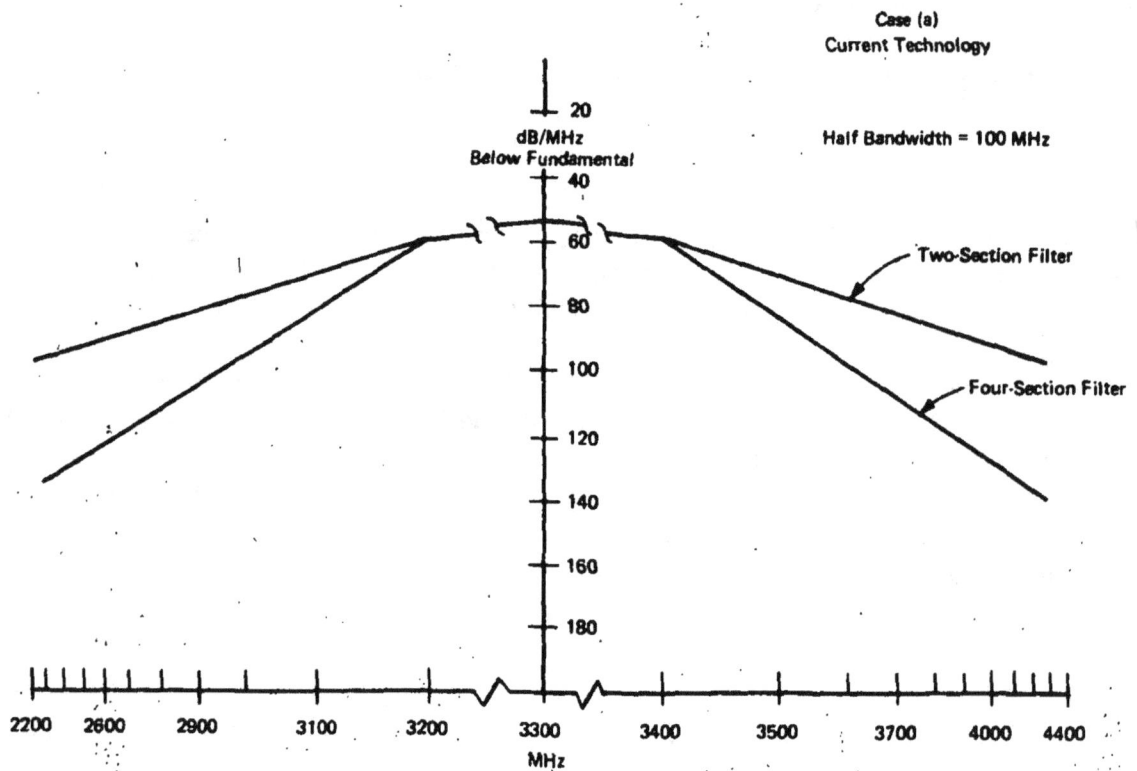

FIGURE 78. — SSPS NOISE ASSOCIATED WITH BASIC DEVICE AND FILTER

Case (b) is also divided into two parts, b2 and b4, which also designate a two- and four-section filter, respectively. However, it is expected that improvement in filter design will enable us to reasonably achieve a half 3 dB width of 50 MHz from the fundamental. This will follow the law of -6 dB per octave (of 50 MHz) per section. Figure 79 shows the straight-line approximation to the performance of these filters. Again the improvement in Amplitron design is reflected in the curves.

The total effective filtering between the antenna and the generator in dBw/MHz is given in Figure 78 and 79 for each of the respective cases, a2, a4, b2, and b4.

d. Noise Power

The system shown in Figure 80 shows the noise power generated in dBw/m² at some distance R from the transmitting antenna which is given by,

$$P \text{ dBw/m}^2 = P_{T_o} - [F - 10 \log \frac{b}{b_o}] + G - 10 \log 4\pi R^2 \tag{3}$$

where

P_{T_o} is the total input power = 98 dBw,
F is the appropriate filtering obtained from Figures 78 and 79,
G is the noise gain of the antenna (from Figure 77),
R is the distance = 381 x 10^7 m = (10 log $4\pi R^2$ = -162.6dB),
b is the bandwidth in MHz, and
b_o = 1 MHz

For most considerations, P_{T_o} will remain the same at 98 dBw and, since the transmitting system is in stationary orbit, the distance R will remain constant, and the equation for P dBw/m² becomes:

$$
\begin{aligned}
P \text{ dBw/m}^2 &= 98 - [F - 10 \log \frac{b}{b_o}] + G - 162.6 \\
&= - [F - 10 \log \frac{b}{b_o}] + G - 64.6
\end{aligned}
\tag{4}
$$

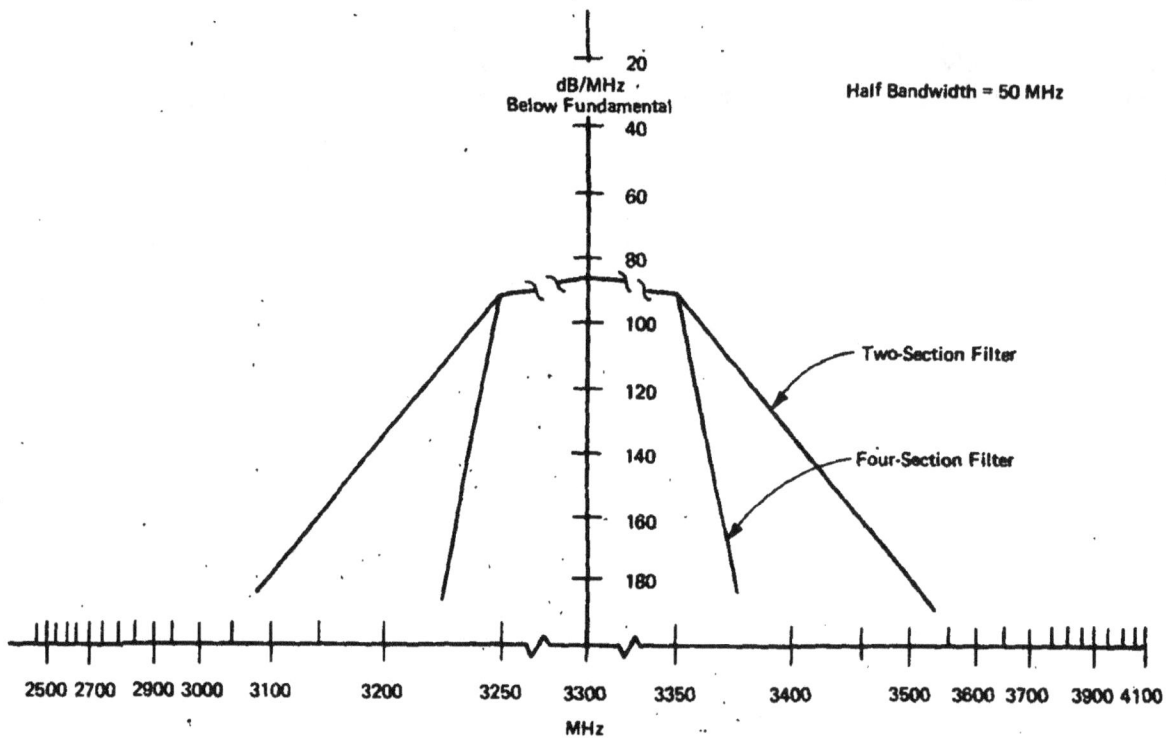

FIGURE 79. — SSPS NOISE ASSOCIATED WITH BASIC DEVICE AND FILTER

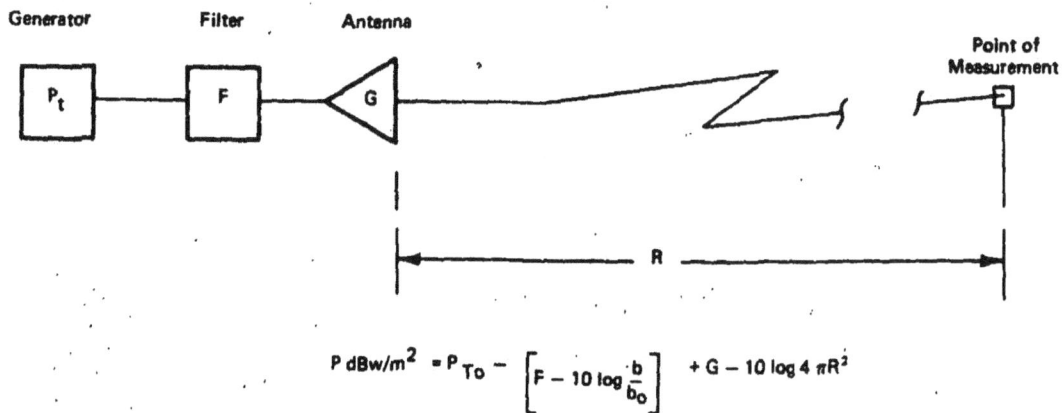

$$P \text{ dBw/m}^2 = P_{To} - \left[F - 10 \log \frac{b}{b_o} \right] + G - 10 \log 4 \pi R^2$$

FIGURE 80. — TRANSMITTING/RECEIVING SYSTEM

The International Radio Regulations limit the power flux density at the Earth's surface from satellites. These regulations are designed to protect radio services on the Earth's surface from destructive interference from satellite transmission in the same frequency bands.

Section 470 NGA of these regulations indicates that, for frequencies between 1670 and 2535 MHz, the power flux density from the SSPS must not exceed -168 dBw in any 4-kHz band. This is to protect fixed services using tropospheric scatter which apparently is the most sensitive system in that band.

From Equation (4):

$$P \text{ dBw/m}^2 = -[F - 10 \log \frac{b}{b_o}] + G - 64.6 = -168 \text{ dBw}$$

$$-[F - 10 \log \frac{b}{b_o}] + G = -103.4$$

From the above equation and Figures 77 and 78, the results for Cases (a2) and (a4) filtering are obtained. It can be seen that tropospheric scatter communication can operate within ± 1500 MHz from the fundamental for two-section filtering and ± 400 MHz for four-section filtering where a single SSPS is deployed (see Figures 81 and 82). An indication of the effect of a large number of SSPS units (100 SSPS) is shown to increase the noise 20 dB. For a four-element filter, the tropo service can operate to within ± 765 MHz of the fundamental.

From Equation (4) and Figures 77 and 79, we obtain the results for Cases (b2) and (b4) filtering. It can be seen that operation is possible within ± 85 MHz for two sections and ± 60 MHz for four sections where a single SSPS is deployed.

The International Radio Regulations also indicate the limit of harmful flux densities in the case of radio astronomy. On page 433 of the regulations, there is a table which indicates that, in the range of the SSPS fundamental frequency, an average level of harmful flux is defined as > -175 dBw/m^2 for an isotropic antenna. Antenna gains in this operation can run to 55 dB as a typical example. This means that the SSPS noise flux density must be kept below -230 dBw/m^2 in order not to interfere. Also in this case, the bandwidth is typically 10 MHz.

From Equation (4) and Figures 78 and 79, we have

$$P \text{ dBw/m}^2 = -[F - 10 \log \frac{b}{b_o}] + G - 64.6 = -230$$

$$-[F - 10 \log \frac{b}{b_o}] + G = -165.4$$

$$10 \log \frac{b}{b_o} = 10$$

106

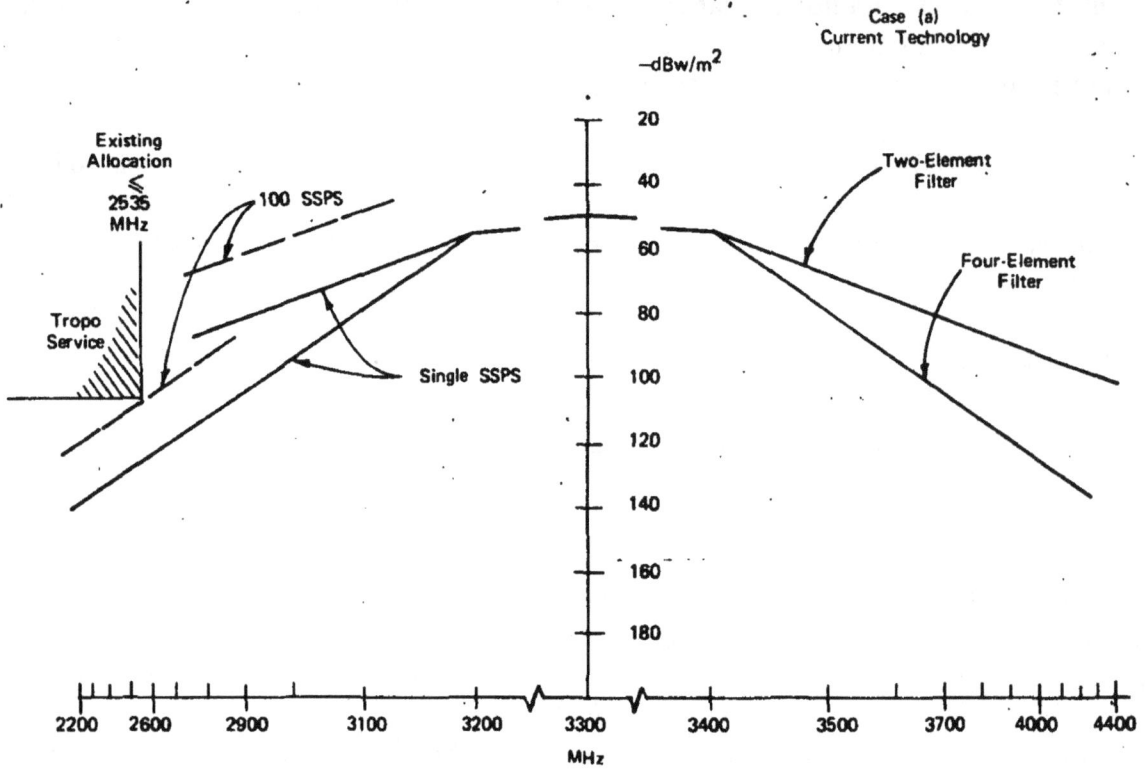

FIGURE 81. — SSPS NOISE ASSOCIATED WITH TROPOSPHERIC SCATTER

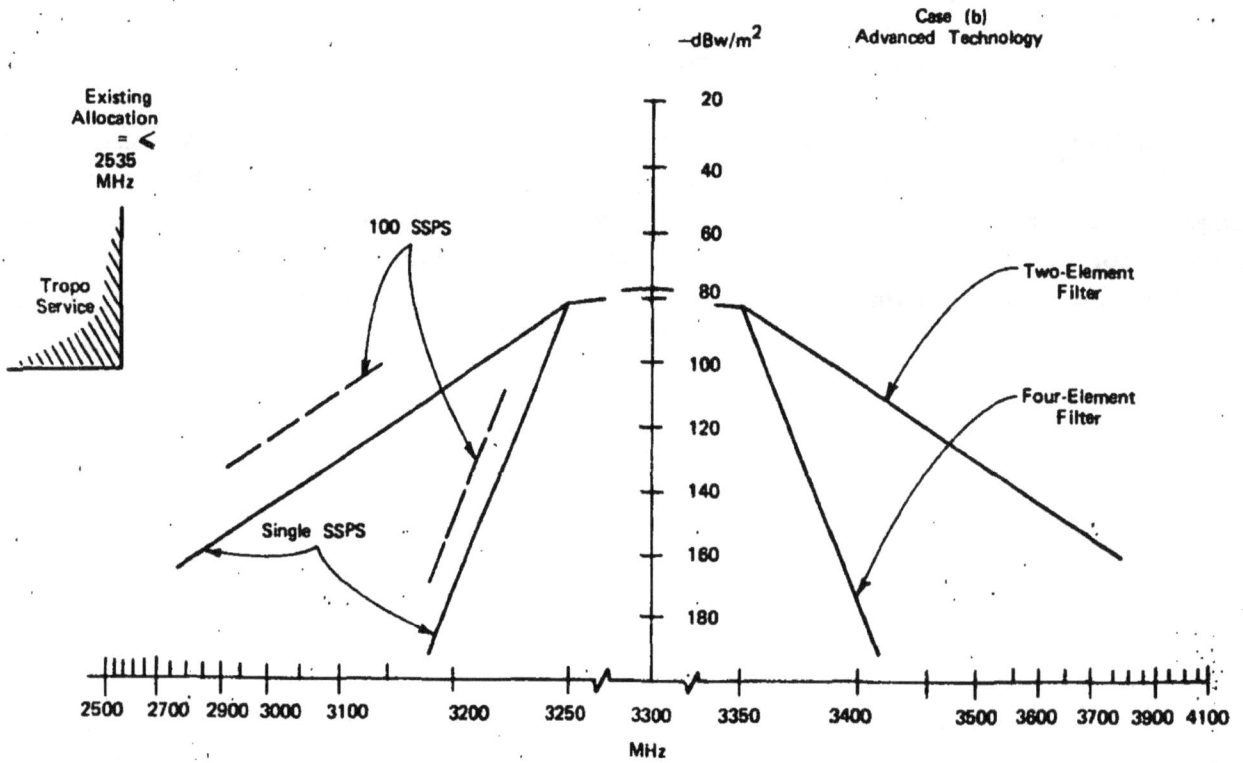

FIGURE 82. — SSPS NOISE ASSOCIATED WITH TROPOSPHERIC SCATTER

107

For Case (a2) we have $\pm 10^5$ MHz

For Case (a4) we have ± 4500 MHz

(See Figure 83.)

For Case (b2) we have ± 500 MHz

For Case (b4) we have ± 160 MHz.

In the above two radio astronomy cases, it was assumed that the radio astronomy antenna was pointing directly at the SSPS. If this antenna is limited to point only to within, say, 2 degrees of the SSPS, the gain of the RA antenna is essentially at the isotropic level. This effectively raises the -165.4 dBw line on Figures 83 and 84 to -110 dB (marked 2 deg).

> With a2 filtering we have ± 8000 MHz
>
> With a4 filtering we have ± 1100 MHz
>
> With b2 filtering we have ± 175 MHz
>
> With b4 filtering we have ± 95 MHz.

e. Typical Radar System

A typical radar system (pages 5-6, Skolnik, Radar Handbook) has the following characteristics:

Noise temperature	$= 838^\circ K$	$= 29.3$ dB
Receiver bandwidth	$= 2$ MHz	$= 63$ dB
Boltzmann's Constant		$= -228.6$ dB
Noise level		$= -136.3$ dB

If we consider a 3.2-meter diameter antenna, the area will be about 8 m² so that the total power per square meter needed to exceed the noise level is $-136.3 - 10 \log 8 = -145.3$ dBw/m². Since radars require a positive S/N ratio to operate, we can let the noise from the SSPS equal -145.3 dBw/m² and still have interference. Therefore, Equation (4) becomes

$$P \text{ dBw/m}^2 = -[F - 10 \log \frac{b}{b_o}] + G - 64.6 = 1453$$

$$-[F - 10 \log \frac{b}{b_o}] + G = -80.7$$

$$10 \log \frac{b}{b_o} = 3.$$

Figure 85 shows the cases for: a2 = ± 1400 MHz
                                                    a4 = ± 460 MHz

Figure 86 shows the cases for: b2 = ± 100 MHz
                                                    b4 = ± 60 MHz

108

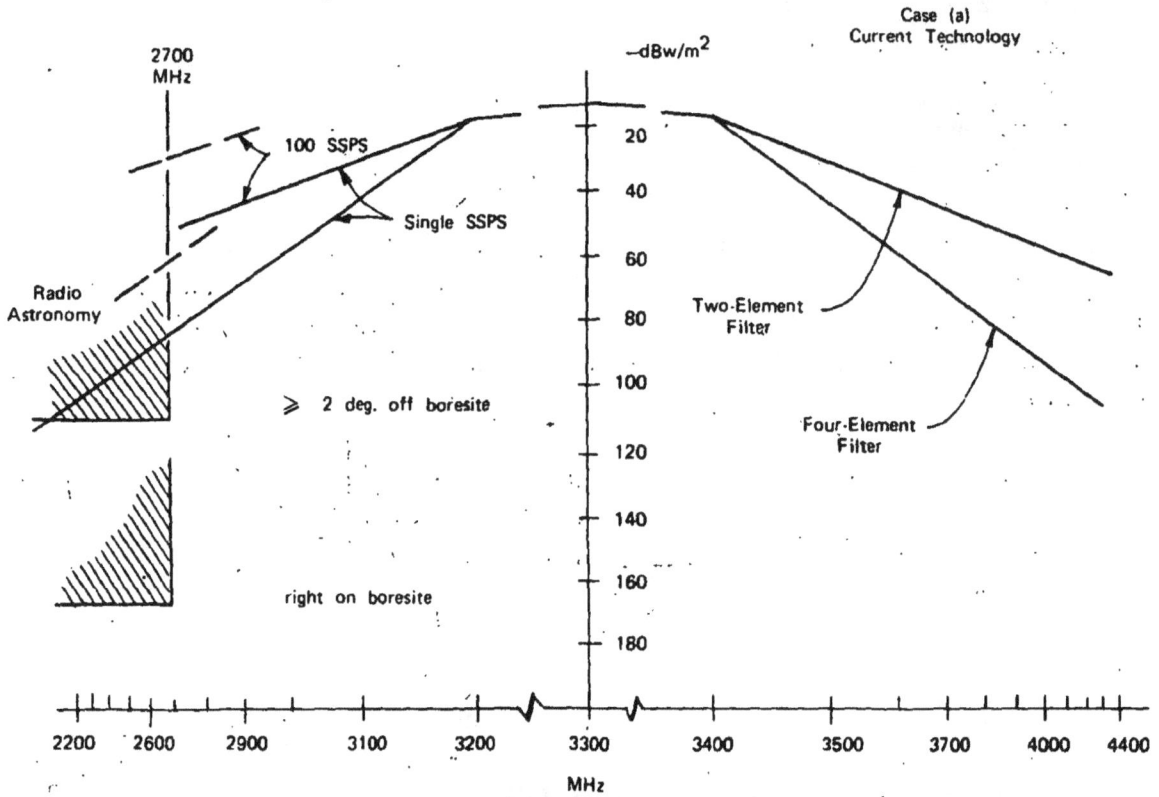

FIGURE 83. — SSPS NOISE ASSOCIATED WITH RADIO ASTRONOMY

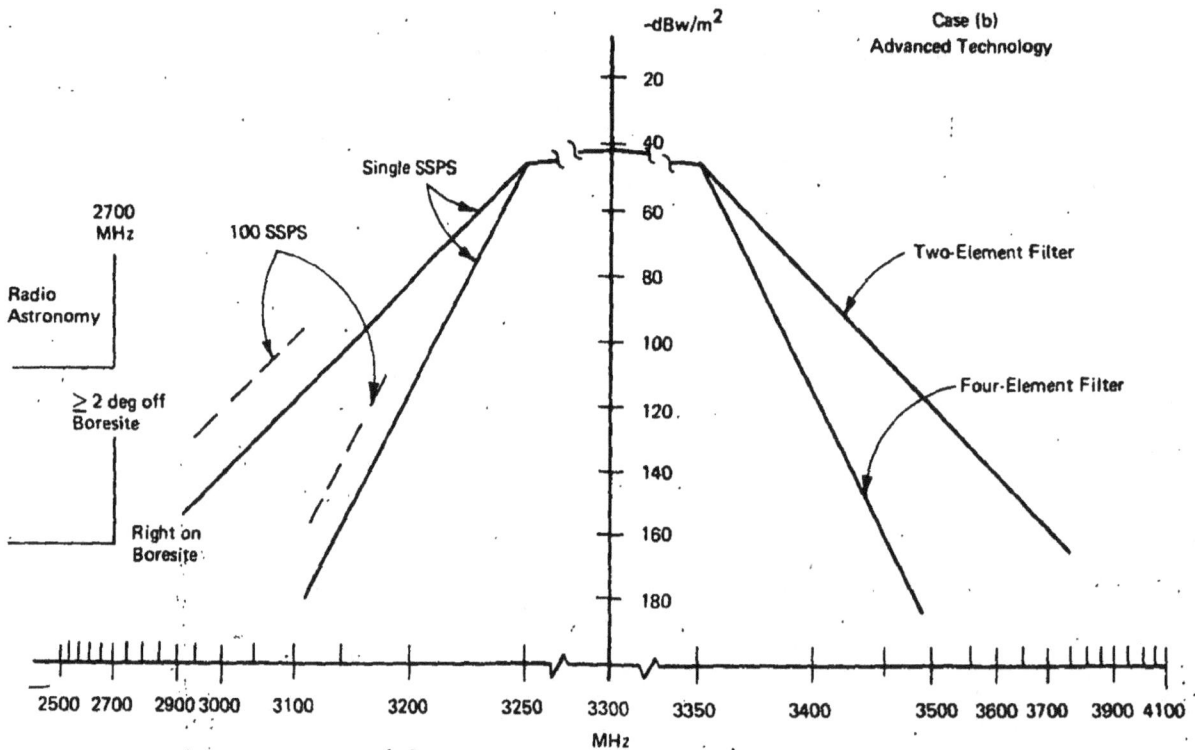

FIGURE 84. — SSPS NOISE ASSOCIATED WITH RADIO ASTRONOMY

109

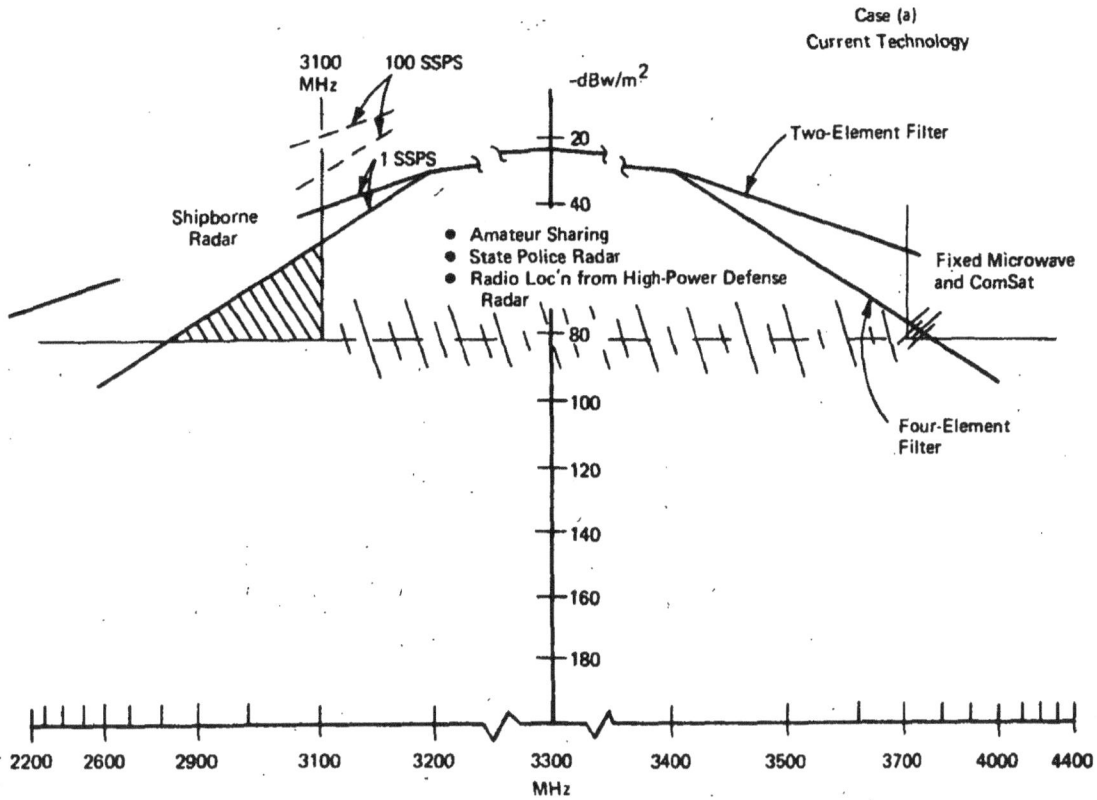

FIGURE 85. — SSPS NOISE ASSOCIATED WITH RADAR (10' Diameter)

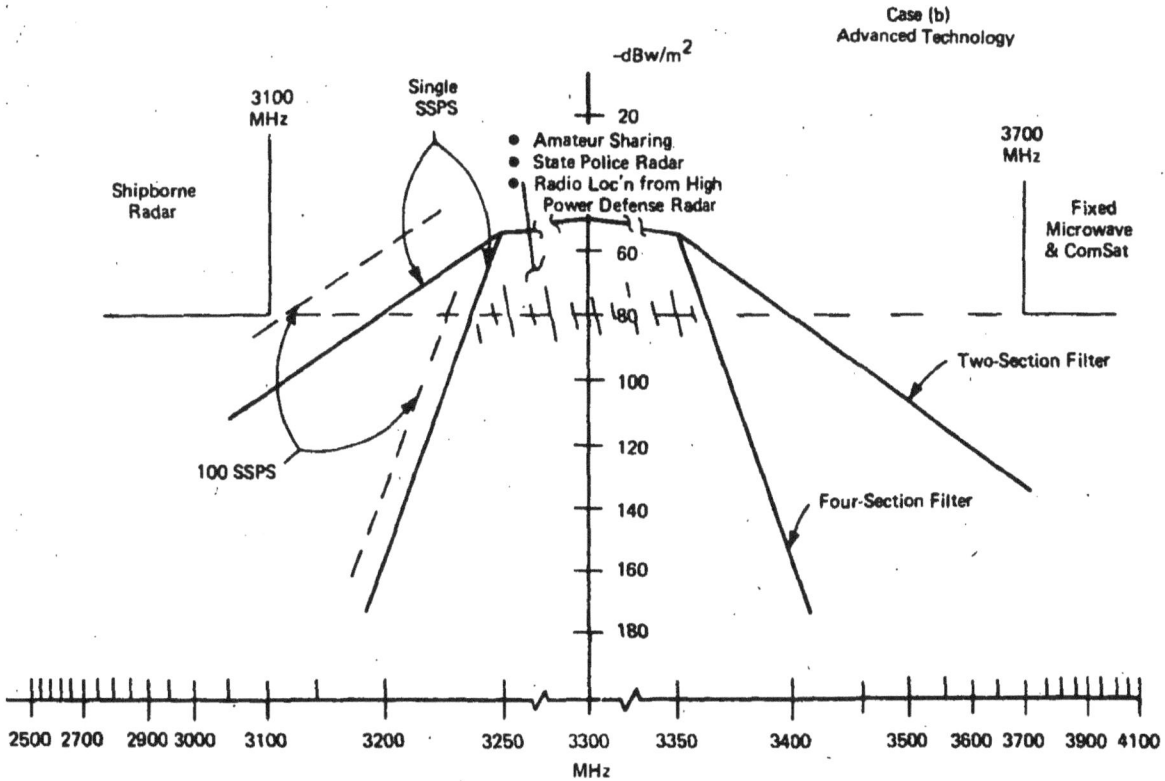

FIGURE 86. — SSPS NOISE ASSOCIATED WITH RADAR (10' Diameter)

110

Table 18 and Figures 77 through 86 show that Amplitron design and filtering techniques are of the utmost importance in minimizing RFI with other users and hence for achieving national and international agreement on frequency allocation for the SSPS as a service.

TABLE 18

SERVICE FREQUENCIES — CASES (a) AND (b)

Type of Service	Case (a)		Case (b)	
	2 Section	4 Section	2 Section	4 Section
Radio Astronomy	$\pm 10^5$ MHz	± 4500 MHz	± 500 MHz	± 160 MHz
with offset	± 8000 MHz	± 1000 MHz	± 175 MHz	± 95 MHz
Tropo	± 1500 MHz	± 400 MHz	± 85 MHz	± 55 MHz
Radar	± 1400 MHz	± 460 MHz	± 100 MHz	± 60 MHz

In summary, there are identifiable approaches which could reduce RFI to internationally acceptable levels. The following listing provides a qualitative overview of RFI.

		3300 MHz Interference
SSPS Fundamental Frequency		
TROPO Service	< 2535 MHz / > 4400 MHz	None
Radio Astronomy (< 2700 MHz)		Slight
Shipborne Radar (< 3100 MHz)		Partial
Fixed Microwave (> 3700 MHz) and Communications		Slight
Amateur Sharing / **State Police Radar** / **Radio Location from** / **High Power Defense Radar**	(3100 MHz to 3700 MHz)	Substantial

NOTE: The data presented here indicating interferences and re-allocations should not be interpreted as being possibly acceptable. Rather they are an identification of small bands of frequency requiring much investigation to determine impact before recommendation as to acceptability could be contemplated.

111

IDENTIFICATION OF KEY ISSUES

Key Technological Issues

<u>Microwave Generation, Transmission, and Rectification.</u> – The following issues are significant for the SSPS development.

a. Technology Investigations

The microwave portion of the electromagnetic spectrum is the most useful for the SSPS power generation, transmission, and rectification. Approaches to generate, transmit, and rectify power in other major segments of the spectrum must be assessed, and the investigation must be documented so that data will be available for national and international discussions and negotiations leading to frequency allocation for the SSPS.

The requirements placed upon the technology of a microwave power system in terms of high efficiency, long life, light-weight, and low cost present significant technical challenges. Fortunately, there are a large number of design resources in the form of systems technology, materials technology, and device technology which can be applied to meet these challenges.

Table 19 summarizes the various design resources which can be used to meet the requirements placed on microwave power transmission in the SSPS.

Major areas of work (Table 20) have been identified which constitute the issues in the generation, transmission, and rectification of microwaves. The table presents a status statement for each area of work as well as its development goal and the basis for assuming what can be achieved. Development goals and technology investigations to achieve the required improvements can be identified.

From a systems point of view, the major item is a detailed investigation of approaches for implementing the adaptive array principle. Such an investigation implies a study of the details of the subarray and the means of establishing the phase reference under steady-state as well as transient conditions. From a device point of view, the major items are related to the performance characteristics of the Amplitron. Many of these performance characteristics are related to the unique requirements of the SSPS application and are not easily derived from terrestrial application experience. The demonstration of a high dc-to-dc microwave system efficiency in the laboratory can be achieved with modest effort to establish the credibility of efficient microwave power transmission.

Specific technological investigations are listed below:

- Demonstrate high efficiency dc-to-dc microwave power transmission,

- Investigate approaches to implementing the adaptive array principle,

TABLE 19

DESIGN RESOURCES TO MEET REQUIREMENTS PLACED ON MICROWAVE POWER TRANSMISSION IN SSPS

Design Resources \ SSPS Requirements	High Efficiency 68%			Passive Radiation of Waste Heat		High Power 5 × 10³ kW		Low Weight in Orbit	Low Cost		Long Life (20 yr)	High Reliability	Ease of System Startup	Low RFI	High Pointing Accuracy	All Weather Capability
	Satellite	Beam	Ground	Satellite	Ground	Satellite	Ground		Satellite	Ground						
Active Phased Array Principle	●			●		●		●	●		◉	●	◉			
Adaptive Array Principle		●						●	●			●	◉	◉	●	
Rectenna Principle			●		●		●			◉	●	●				●
Selection of Wavelength	◉		◉	◉	◉	◉	◉	◉	◉	◉	●	●		◉		●
Amplitron Principle	●			◉		◉		●	●		●	●	◉			
High Magnetic Field	●			◉		◉		●	●							
Sm-Co Magnet Material	◉			◉		◉		◉	◉							
Output Power Sensing Variable Magnetic Field		●										●	●	◉	●	
Pure Metal Secondary Emitting Cathode	△							△	△		●	●	●			
Pyrolytic Graphite Radiator				◉		●		●	●		◉	◉				
High Operating Temperature	△		△	◉	△	●		●	●		△	△				
Small Individual Tube Size				●		◉		●	●			◉				
GaAs Schottky-Barrier Diode				△	●	◉	●	●	●	●	●	●	●			
Microwave Interferometer Principle													●		●	
Auxiliary Filters	△		△	△	△	◉	◉	△	△	△				●		
Subarray Principle						●		●	●							
Laser Beam								●	△						●	
Lightweight, High Strength Materials								●							◉	

● Strong Positive Correlation   ◉ Weak Positive Correlation   △ Weak Negative Correlation

TABLE 20

MICROWAVE POWER GENERATION, TRANSMISSION, AND RECTIFICATION ISSUES

Major Areas of Work	Status	Development Goal	Basis for Improvement
D.C. Power Transmission, and Distribution	Normal	Normal	Normal*
Microwave Beam Approach	D.C. on orbit/D.C. on ground = 54% (expected with current technology)	70%	Design optimization
Transmitting Antenna Assembly	Early concept	Passive control of structure Electronic control of phase front	Integration of known technologies
First Level of Integrated Assembly (Subarray)	Schemes available	Fundamental building block	Integration of microwave conversion devices, filters, and waveguides
D.C. to Microwave Conversion (Device)	Microwave/DC = 85% RFI high	90% noise and harmonics Compatible with 60-kV operation	Design optimization and incorporation of filter
Local Collection Rectification & Filtering	DC/Microwave rectification = 75%	90%	Design optimization and materials selection
Physical Arrangement of Rectenna Elements	Normal*	Normal*	Normal*
Integration of Switchable Assemblies	Normal*	Normal*	Normal*
Interface with User Network	Normal*	Normal*	Normal*
Local & Network Power Distribution, Cond. & Ctl.	Normal*	Normal*	Normal*
Ground-Based Flight System Monitoring & Control	Active ground monitoring and control	Acquisition establishment and control of phase front	Open circuit before Amplitron, if necessary
Site Definition	Early concept	High ecological compatibility	Advanced planning for sites and distrib'n grids

*Normal – No extraordinary information is identifiable at this time. (See Table 21).

114

TABLE 20 (Continued)

Major Areas of Work	Status	Development Goal	Basis for Improvement
National & International Agency Actions	Current plans known. RFI potential high	Obtain frequency allocation	Early development and ground-work for frequency allocation
Band of Frequency for Power Transmission	Microwave	Try to identify advantages & penalties to other band	None known but vital for negotiation & frequency allocation
Specific Near-Optimum Frequency for Power Transmission	3.3 GHz	Identify advantages & disadvantages of neighboring frequencies	Understanding of hardware parameters, ionospheric & tropospheric losses
Select Orbital Locations	Early station at stable point $\doteq$ 123°W (prelim.)	Determine how to make most effective use of prime locations	Early planning analysis & test will lead to the understanding of the nature of the on-orbit vehicle behaviors.
Select Ground Locations	In currently to-be-developed areas. Early receiving antenna Site – SW desert of U.S.	Take ecological advantage of large multiple control power stations with connecting grid approach	Advantage afforded by long-range planning.
Define Interference Effects & Present Solutions	55db/MHz below fundamental	75dB/MHz for ±25 MHz 100 dB/MHz outside ±200 MHz	Tube and filtering design optimization
Arrive at Criteria for Policy Decisions	To be based primarily on importance of SSPS power	Determine approach to minimize impact on other users near fundamental & at harmonics	Ecological advantage of SSPS power compared to ecological disadvantage to electromagnetic spectrum allocation & displacement of users
Microwave Biological Effects	Within acceptable protected work area limits. Outside guard band, within national standards	Establish long-term effects & protection req'ts outside normal guard band, establish effects & procedures inside guard bands, particularly for birds	No currently known adverse effects within the national standards
Environmental Effects	Preliminary investigations. 10% inefficiency	Assure that atmospheric & other environmental effects are acceptable	Low heat generated in process of providing power

TABLE 21

POWER CONVERSION TRANSMISSION, RECEPTION, AND CONTROL
(NORMAL)

Status:

- In general, physics is understood and qualitative requirements are known.
- Several approaches required to be developed and assessed.
- Engineering, production, and verification required.
- All elements are in the analytical feasibility assessment phase.
- A plan has been developed.
- Ionospheric, atmospheric, and other transmission efficiencies are expected to be 87%.
- Stated efficiencies are those to be expected with present technology.

Development Goal:

- Producibility, low cost, high reliability and long life (approx. 30 years required).
- Low mass and high density transport for flight items.
- High integration of structural and power distribution technologies.
- No new technologies identified unless noted.

Basis for Improvement:

- Few different parts at very high production rates operating in hard vacuum at high voltage (20,000V) and low current.

Required Action (General):

- Determine near optimum design approach and develop highly efficient processes for manufacture, delivery, assembly, and operation.
- Proceed in ordered and directed steps in an assessment, technology, design, verification and demonstration program implementing systems requirements.
- Provide for frequent Go/Hold/No-Go decision points.
- Normal identification and evaluation of design approaches required for items on first page of Table 20 that near optimum system design is achieved in terms of high performance and low cost with programmatic confidence.
- Normal determination of ecological impacts and development of recommended solutions for items on third page of Table 20 through (c) 3.0.
- Proceed with plan.

116

Design, construct, and evaluate the Amplitron device with associated filtering based on criteria which include efficiency, noise and harmonics output, start-up behavior at very cold temperatures and in ultra-high vacuum, operation using variable magnetic fields features, long operating life, low cost, producibility, and maintainability.

The above issues present a broad outline of the microwave power generation, transmission and rectification development program. Subsidiary issues in each of the work areas will be identified as development proceeds and specific designs are analyzed.

b. Biological Effects of Microwaves

Standards for microwave exposures ranging from a continuous exposure limit of 10 mW/cm^2 in the United States to 0.01 mW/cm^2 in the Soviet Union have been established. In the United States, the standard is based on microwave heating of body tissues, while Soviet investigators believe that the central nervous system is affected by microwaves, even at very low exposure levels. In view of the different interpretations of the effects of microwave exposure, there is a need to obtain a better understanding of these effects and to develop experimental procedures to assure that the by-products of microwave-generating equipment, such as X-rays, ozone, and oxides of nitrogen, in addition to extraneous environmental conditions imposed on laboratory test animals would not lead to a misinterpretation of the laboratory observations.

The lack of widely accepted standards for microwave exposure requires that the SSPS design be able to accommodate a range of power flux densities. An understanding of the specific SSPS-induced environment, predictions, analyses and measurements will have to be an essential part of the development program as additional information on microwave biological effects is made available.

Precise control of the microwave beam through SSPS stabilization and automatic phase control will assure that the microwave power will be efficiently transmitted to the receiving antenna. The transmitting antenna size, the shape of the microwave power distribution across the antenna, and the total power transmitted will determine the level of microwave power flux densities in the beam reaching the Earth. The SSPS fail-safe design features will include a pilot signal transmitted from the ground to the SSPS to assure precise beam-pointing and remotely operated switches to open solar cell array circuits and shut off all power to the microwave generators.

There is a wide choice of the microwave power flux densities incident upon the receiving antenna at a desired total microwave power output. The variations in microwave power flux densities can be obtained by adjusting the diameters of the transmitting and receiving antennas. A guard ring can be provided so that the level of microwave exposure of the public will be less substantial than 0.1 mW/cm^2, because there is considerable design flexibility so that the SSPS can be operated to meet agreed-upon standards for microwave exposure. Figure 87 shows the microwave power density distribution for very simple power distributions across the transmitting antenna.

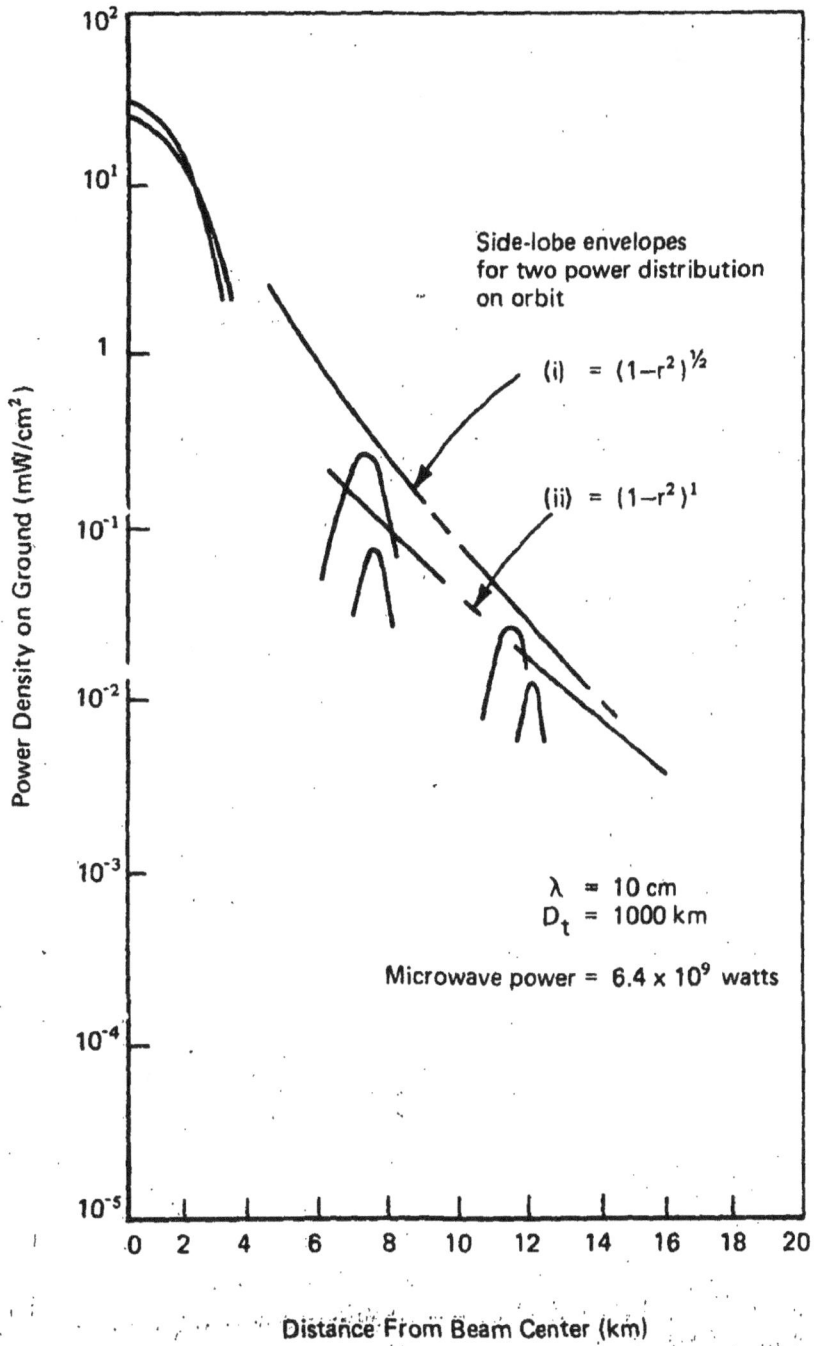

FIGURE 87. — MICROWAVE POWER DENSITY DISTRIBUTION ON GROUND

118

This figure shows the microwave power density for the power distribution discussed in the last section of the previous chapter (Effects of SSPS RFI on Other Users), and Table 17 which will result in the formation of side lobes. It indicates a continuing decrease with distance from the beam center; e.g., at about 15 km, it is less than 0.01 mW/cm². This is the lowest value set as a limit for microwave exposure by any country. The permissible exposure value for humans or other forms of life by the time the SSPS becomes operational has yet to be determined on an internationally agreed-upon basis.

In addition to these effects, interference with electronic equipment, medical instrumentation, or electro-explosive devices must be precluded as well. Their sensitivity to a low level of microwave exposure will have to be established. To realize this objective, industrywide standards may be necessary.

The effects on birds flying through the beam is not known. Research on the effects of microwaves on birds at the level to be encountered in the microwave beam will have to be carried out. Preliminary evidence indicates that birds can be affected at levels of microwave exposure of 25 to 40 mW/cm² in the X-band.

The effects of microwave exposure on aircraft flying through the beam must also be considered. The shielding effects of the metal fuselage and the very short time of flight through the beam would preclude significant exposure to humans. Protection of aircraft fuel tanks from electrical discharges is a standard design feature, but the absence of microwave-induced effects will have to be confirmed. In addition, interference with aircraft communication and radar equipment will have to be precluded.

Because of the concerns with potentially adverse effects being claimed to result even at low microwave power density, it is vital that the microwave system design requirements of the SSPS lead to the demonstration of low-exposure levels so that safe operations accepted on an international basis will result.

The major elements of a microwave biological effects R&D program to support the SSPS development are listed below:

- Limit R&D to SSPS specifics

  Frequency: 3.3 GHz cw
  Power flux density on ground: 30 mW/cm² in central part of beam
                                       0.01 mW/cm² beyond a radius of 15 km
  On orbit: 25 kW/m² in central portion of beam

- Determine whether or not the near-optimum frequency and power flux densities for SSPS as derived for other considerations should be changed;

119

- Develop understanding of specific SSPS-induced environment;

- Develop understanding of effects on birds and other animals anticipated to be exposed;

- Develop required procedures and monitoring techniques;

- Make clear any penalties to be faced for operating at the near-optimum frequency and flux density; and

- Determine other penalties associated with operating at frequency and flux density levels which deviate from the near-optimum values.

   c. National and International Negotiations

The prerequisites for frequency allocation negotiations at the national and international levels are as follows:

- Determine the potential of the SSPS to meet future national and international electrical power needs;

- Determine detailed impact on SSPS if other frequencies are allocated;

- Determine detailed impact on other users of the electromagnetic spectrum; and

- Analyze secondary effects, including effects of operational transients, scattering by accidental aircraft intrusion in beam and deliberate interference with transmission by the release of scattering media in beam.

The ease with which negotiations can be accomplished will depend primarily on how critical the power need is and on how much of the need can be fulfilled by the SSPS. The impact on other users, of course, could be minimized by obtaining a frequency allocation early so that only a limited number of systems would be developed that would have to be modified or redesigned completely once the SSPS became operational. It is then in the nation's general interest to establish whether or not a potential role for the SSPS may exist and, if so, to initiate investigations and discussions to elicit the issues to be resolved in allocation negotiations. The earlier this process is started, the less difficult and trying it will be to all concerned.

Solar Energy Conversion. — The key issues which will require further investigation time and the performance goal for each are listed below:

Key Issue	Goal
1. Solar cell performance	18% efficiency, 2-mil-thick cell
2. Solar cell cost	0.38 cent per cm^2
3. Blanket weight	950 W/kg (430 W/lb)
4. Blanket cost	0.68 cent per cm^2
5. Concentration optimization	Array: $310/kW, 3 lb/kW
6. Long life	30-year-life, 6% degradation in 5 years
7. Energy input to process	1-3 year payback
8. High-voltage circuit control	40 kV − 5% loss
9. Operation	Automatic assembly and replacement

Within each of the key issues there are distinct work areas which will precipitate different degrees of urgency, depending on the specific objective of the effort and its relationship to the final system configuration. Table 22 presents the areas which have been identified. The table is divided into four groups, the first of which includes those areas which require immediate action to maximize the probability of technology readiness at the time of production. The fourth group represents those areas which have the least immediate requirement at the present time, but can still be identified as areas for future work. Investigations carried out without a priority framework in terms of SSPS development objectives are unlikely to result in an effective and coordinated effort. The relative importance of each of the work areas as they relate to one another, as well as to the system concept, has to be related to the overall SSPS development program.

With each of the items listed under the specific work areas is a reference to the key issue number. The objective of the particular work area is also listed and the importance of that effort to the total system is indicated.

Following the table, each key factor is described in detail, the research objectives are listed, and the approach to be pursued is indicated.

    a. Key Issue No. 1 — Solar Cell Performance Improvement

Investigations into methods of improving solar cell efficiency are extremely important to the weight and size reductions required for SSPS. The efficiency must be increased from about 14% to 18%, while at the same time reducing the thickness of the devices from about 250 to 50 μm. The overall task is expected to require 10 years. The first three or four years will be concerned primarily with theoretical and laboratory studies of potential efficiency-improved techniques and production processes. The approach will be to pursue:

1. The use of low resistivity silicon,

2. Investigations into the theoretical and experimental development of new conversion devices and alternate materials,

121

TABLE 22

KEY STUDY AREAS IN SOLAR ENERGY CONVERSION

GROUP I

Item	Issue No.	Work Areas	Objective
1	1	Cell Quantum Efficiency Increase	(I_{SC} 34→42mA/cm^2)
2	2	Cell Fabrication Cost Reduction	($1.00/cm^2→0.01/cm^2)
3	3	Cell Thickness (Weight) Reduction	(8 mil → 2-mil)
4	3	Ultra Light-weight Blanket Design	(70→430 W/lb)
5	5	Concentration Ratio Optimization	(N = 1→N = 4)
6	5	Concentrator Filter Design	(Minimize temperature)
7	6	High-Voltage Plasma Protection	(40 kV capability)
8	8	High-Voltage Switching	(40 kV capability)

GROUP II

Item	Issue No.	Work Areas	Objective
1	1	Cell Voltage Increase	(V_{OC} 540→780mV)
2	2	Semiconductor Material Cost Reduction	(12¢/cm^2→0.2¢/cm^2)
3	4	Blanket Cost Reduction	($2.50/cm^2→$0.015/cm^2)
4	6	Improved Radiation Resistance	(12% deg. in 2 years→6% in 5 years)
5	6	Reduction of Thermal Cycle Fatigue	(5 years→30 years life)
6	8	Optimized Cell Layout and Bus Design	(Use cells for busing)

GROUP III

Item	Issue No.	Work Areas	Objective
1	1	Cell Design Optimization	(GaAs, VMJ, other)
2	5	Concentrator Cost Reduction	(1/10 of Blanket Cost)
3	6	Minimized Meteorite Loss	
4	8	Minimized Magnetic Moments	(Layout with balance)
5	8	Minimized Busing Losses	(Structure used as bus)
6	8	High Level DC Power Distribution	(Conventional versus superconductor)

GROUP IV

Item	Issue No.	Work Areas	Objective
1	7	Reduced Energy Requirements of Processes	(< 1 year energy pay back)
2	8	Power Control and Circuit Protection	(Shut down and repair)
3	9	In Orbit Assembly Concepts for Array	(Mechanized assembly)
4	9	In Orbit Repair or Replacement of Array	(Easy replacement)
5	9	Magnetic Attitude Control	(Layout control circuits)

3. The development of improved production processes,

4. Improvements in collection efficiency, and optimization of temperatures reached with solar concentrators,

5. The use of multijunction cell designs,

6. The use of lithium doping,

7. Improved antireflection coatings,

8. Methods of decreasing surface recombination velocity,

9. Methods of improving high-temperature and high solar intensity performance, and

10. New methods for environmentally testing devices to provide more reliable and better designed parts.

   b. Key Issue No. 2 — Solar Cell Cost Reduction

The need for reducing the cost of solar cells is a critical factor for the SSPS and has been recognized as the prime item not only for SSPS, but also for terrestrial applications of photovoltaic energy conversion techniques. Solar cells used in the space program presently cost about $80 per watt, while the SSPS requires a cost of about $0.40 per watt. Although it is known that a major portion of this reduction in cost will be possible as a result of mass production, there still are several approaches that need to be studied. Solar cell cost reduction will be accomplished by developing:

1. New silicon material sources,

2. New crystal-growing processes,

3. New solar cell fabrication processes,

4. Ways of minimizing the heat or power input to the processes,

5. New mass production and automated processing techniques.

   c. Key Issue No. 3 — Solar Cell Array Blanket Improvement

The SSPS will utilize very large area solar cell arrays that will be able to be effectively handled, only if large integrated submodules or blankets of cells can be developed. Presently, solar cell arrays are made much like an art mosaic where individual cells are fitted, interconnected, and bonded to substrates. Power-to-weight ratios of about 60 W/lb could presently be achieved, but ultra light-weight blankets of over 400 W/lb are required for the SSPS. This can be accomplished by developing:

1. Arrays utilizing only the minimum thickness cell, cover, and substrate;

2. New blanket fabrication techniques;

3. Light-weight blanket-attachment techniques;

4. New, light-weight, structural weaving techniques;

5. Improved radiation-resistant materials;

6. New thermal control coatings;

7. Designs capable of surviving extreme thermal cycling; and a

8. New method for environmentally testing blankets to provide more reliable and better designed arrays.

d. Key Issue No. 4 — Solar Cell Array Blanket Cost Reduction

Cost of the SSPS is probably the most critical factor. In the case of the array blanket, it is important that additional costs beyond the cost of the solar cell itself be kept to a very low value. The interconnection of the cells and the encapsulation in a blanket must therefore be reduced from the present level of about $180 per watt (including cell costs) to a level of about $0.60 per watt. The approach to reducing these costs is to develop:

1. New and simpler blanket fabrication processes,

2. New continuous laminating techniques,

3. New printed circuit interconnector techniques, and

4. New mass production techniques that allow mechanization and automation.

e. Key Issue No. 5 — Solar Cell Concentration Technique Improvement

To achieve the low cost requirement of the SSPS, it will be necessary to utilize solar energy concentration. Concentration makes it possible to generate more power from each solar cell and, if the concentration mechanism is much less expensive than the solar cell, a significant cost reduction can be achieved. The SSPS array analysis showed that the use of a concentration ratio of 3 reduced the array costs from $1.80 per watt to $0.90 per watt. The approach to achieving this technology is to develop:

1. Light-weight mirror-design concepts,

2. Accurate thermal analysis and control,

3. New filter designs to control temperature,

4. Technology for coating large area plastic panels with filters and mirrors,

5. Light-weight structures to support and control orientation of mirrors, and

6. Low-cost filter/mirror structures.

   f. Key Issue No. 6 -- Long Life of Solar Array in Space Environment

The goal of the SSPS is to produce power at high voltage in a relatively stable thermal environment (except during predictable eclipses) over a 30-year period. The cell must be extremely light in weight, yet afford protection from the space environment. The exposure of the solar blanket to the ultraviolet radiation, as well as the particulate radiation, will require protection to ensure long life. The objective is 6% degradation over 5 years. The thermal cycling of the solar blanket during the eclipse period must be accounted for in the system and cell design. The method of cell interconnect and bussing will determine the local thermally induced stresses and ultimately determine the mean time to failure due to fatigue. Techniques to reduce the environmental effects include:

1. Improved radiation-resistant materials,

2. Radiation spectral tailoring to minimize unusable solar radiation,

3. High-voltage plasma protection,

4. Low internal electrical resistance to allow higher current flow,

5. Circuit optimization to balance high voltage versus magnetic moment,

6. Material selection to prevent thermal cycling fatigue,

7. Selection of absorption and emission coefficients for optimal thermal control, and

8. Meteorite hardening optimized with weight and power loss considered.

   g. Key Issue No. 7 - Minimization of Energy Input to Processes

Preliminary investigations of the amount of energy utilized in producing silicon solar cell arrays by present processes indicate that it would take the SSPS several years of operation to return the energy used to make the array. Although the energy is supplied directly by both chemical (coal) and electrical sources, the electrical energy requirements are of primary concern. The objective is to

125

reduce the process electrical energy requirements to a low level, so that about three months of operation of the SSPS will generate the amount of energy utilized. This can be accomplished by developing new processes and new equipment that will provide energy efficiencies in the following areas:

1. Production of metallurgical-grade silicon,

2. Production of high-purity silicon metal,

3. Formation of large area silicon crystals,

4. Formation of P-N junctions in cells,

5. Attachment of metal contacts on cells,

6. Deposition of antireflection coatings,

7. Interconnection techniques,

8. Blanket-lamination techniques,

9. Concentrator mirror formation,

10. Concentrator filter deposition, and

11. Selection of materials used in all of these areas.

The energy requirements for all the conventional processes have to be assessed to determine areas where substantial reductions in energy use will be significant. Theoretical energy-requirement limits for performing each reaction or operation can then be determined so that the potential and objectives are known. New processes and equipment will then be designed that meet these objectives.

h. Key Issue No. 8 — High-Voltage, High-Power Switching

Multi-megawatt solar power generation will require switching protection and control systems capable of reliable operation at a high level. The objective of this program is to determine the best type or types of switching devices for this application and to develop and test representative modules. Specific objectives are:

1. A switch capable of connecting a solar panel submodule of nominally 40,000 volts and 500-ampere capacity to a 40,000 V-dc collector bus; with potential to achieve 60,000 V-dc.

126

2. A protective device (fuse or circuit breaker) suitable for protecting a 500-amp subcircuit connected to a 100,000-amp, 40,000-volt bus.

Both devices must perform reliably in the environment of Earth synchronous orbit with a MTBF of 10 years. Design must be fail-safe. Weight of a switch and protective device combined will not exceed 3 kg. The switch will be tailored to solar panel characteristics (e.g., high open-circuit voltage when cold). Types of switching to be investigated will include mechanical switches, solid-state devices (SCR's, bipolar transistor, field-effect transistors, etc.) and plasma devices (vacuum tubes, thyratrons). Control signals will be low level, available for centralized computer control, isolated from the high power circuit and fail-safe.

i. Key Issue No. 9 – High Voltage Circuit Design

The generation of high currents induce magnetic moments which can react with the natural magnetic environment and cause torques upon the SSPS or result in internal stresses caused by the interaction "self induced" local magnetic fields. The high voltage also could lead to corona formation or other ionized gas phenomena which could reduce the life of the component. The bussing of the high currents, in addition to the magnetic effects, also has an internal resistance associated with them. By judiciously sizing the current-carrying busses and using the structure for bussing when possible, the power losses can be optimized from a weight point of view. The solar cells can be sized to allow extremely long circuits with the necessary parallel members to eliminate the majority of the cross busses as well as the internal busses in the SSPS structure. The solar cells can accomplish the majority of the current-carrying function while at the same time generating new power. A trade-off must be investigated on the relative magnetic moment that can be accepted versus the spacing between high electrical potential current elements under the particular environment.

j. Key Issue No. 10 – High Level dc Power Distribution

Present high-power satellite electrical systems are in the range of a few kilowatts at voltages up to 100. Transmission distances are short. Future space power systems in the multi-megawatt range will require efficient transmission lines over considerable distances at minimum weight. The objectives are to determine:

1. The optimum high-power transmission line which can be built of conventional material (aluminum). Initial design will be for 40,000 volts, 100,000 amps, over a distance of 3 miles. Conductors will be self-supporting and entire structures could be used as structural elements of spacecraft (e.g., to support a solar collector panel). Weight and cost must be minimized.

2. The trade-off between possible approaches based on ease of assembly, cost, weight, reliability (including cost of maintenance), electrical efficiency, and effect on associated systems (e.g., reduction in solar panel weight by using transmission line as a structural support).

The minimum MTBF for the system chosen will be 10 years with appropriate preventive maintenance procedures. A failure mode must either be non-destructive both to the line itself and to associated systems. A monitoring system must be provided allowing sufficient time to shut down without damage.

k. Key Issue No. 11 — Power Control and Circuit Protection for an 8,000-MW Solar Cell Array

The technology needed for the power control and circuit protection of an 8,000-MW solar cell array must be established. The arrays are assumed to be divided into sections small enough to be taken off line without disruption of operation of the remaining sections. A philosophy of protection circuits can be developed for use as a general guideline in evaluating candidate protection methods. The protection circuits and control logic to be developed will, as a minimum, provide for:

1. Protection from solar panel overvoltage resulting after a period in the Earth's shadow,

2. Programmed emergency shutdown to minimize component damage,

3. Detection and automatic isolation of faults to limit failure propagation, and

4. Manual (ground station-initiated) control as well as automatic control.

Analog and other simulations to demonstrate the logic effectiveness will be required. The demonstrations will apply to a portion of the solar array large enough to be deemed significant and capable of being extrapolated to a full-size array. The requirements for telemetry, type, and placement of sensors for protection purposes have to be established. A cost analysis and trade-offs of a number of protection circuit concepts to establish a cost-effective approach will be required.

ℓ. Key Issue No. 12 — Solar Array Assembly in Orbit

The SSPS will require very large area solar cell arrays that will have to be transported to orbit in sections. Therefore, simple assembly techniques will have to be developed for connecting small modules both mechanically and electrically. The objective will be to design the small modules so that they can be deployed and connected by either automatic mechanisms or by manned (tele-operator) systems. This will be accomplished by developing:

1. Solar cell blankets that can be rolled or folded,

2. Blankets that are standardized,

3. Blankets that have electrical terminals that can be connected in orbit to other terminals by simple techniques such as pop rivets, laser welds, magnetic clips, etc.,

4. Blankets that have mechanical support attachment points built in so that they can be simply deployed in orbit within the basic structure of the array, and

5. Blankets that have mechanical and electrical attachment techniques that allow in-orbit repair or replacement.

Earth-to-Orbit Transportation. —

a. Ground to Low-Earth Orbit Transportation

The space shuttle now under development provides the necessary first steps toward a very high-volume, low-cost, Earth-to-orbit transportation system. This shuttle can be used for SSPS technology verification and flight demonstration activities and for transporting elements of a prototype SSPS into LEO. Operational experience with the prototype SSPS will be essential to permit an orderly evolution to the very-high-volume Earth-to-orbit-to-synchronous transportation system needed for an operational SSPS. The basic differences between the current space shuttle and the high-volume-traffic SSPS transportation system are as follows:

Requirement	Current EOS	SSPS EOS
Flights/year	30–50	>500
Transportation cost goal	Order of magnitude reduction in cost of transportation to orbit	Low operating costs
Payload density	High	Low–medium
Payload recovery	Yes	No
Payload to LEO	65K	To be determined
Upper-stage compatibility	Chemical tug	Ion tug

Figure 88 indicates the rationale for identifying advanced Earth-to-orbit shuttle requirements for transporting the SSPS and supporting systems to an appropriate synchronous orbit altitude. The nominal building block sizes for the solar cell arrays and microwave antenna element payloads would be established from representative SSPS configurations. Nominal component sizes and orbital assembly factors inherent in the large orbital configuration can then be identified. These SSPS payload sizing requirements and orbital assembly considerations could then be combined with an appropriate LEO-to-synchronous orbit transportation mode, including the most desirable propulsion elements, to identify the combined total payload size with which an advanced shuttle must contend so that a desirable low-operating-cost transportation system could be evolved.

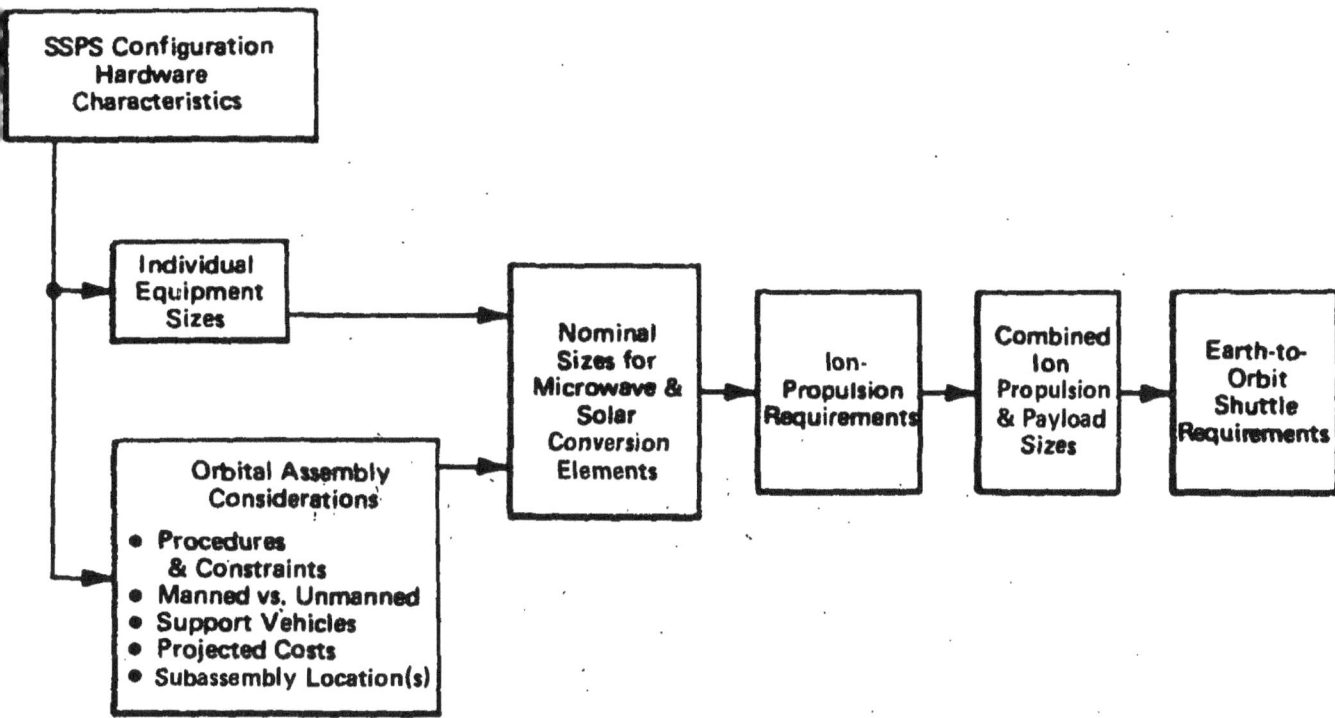

FIGURE 88   ADVANCED SHUTTLE REQUIREMENTS DEFINITION

b. LEO-to-Synchronous Orbit Transportation

The LEO-to-synchronous orbit transportation system under consideration by NASA includes reusable space tugs using cryogenic propellants or storable propellants. None of these chemically powered upper-stage systems has sufficiently high performance for an SSPS. For an SSPS it presently appears necessary to develop advanced high-performance propulsion systems for exclusive operation in the space environment.

Ion propulsion systems are the most likely choice for LEO-to-synchronous orbit transportation systems. A comparison of current ion system technology and that required for an SSPS follows:

ION PROPULSION SYSTEM TECHNOLOGY

Characteristic	Current Value	SSPS Value
Overall specific weight, lb m/kW	100–150	15
power system	50–100	5
thruster system	~50	10
Overall system efficiency	>0.7	0.7
Propulsion Time (thruster life), hr	>8000 ground test >3500 in space	1 yr ~9000 hr
Specific impulse, sec	20,000 demonstrated	8000
Thruster design		
diameter, cm	30–150*	30–150
thrust, lb	0.03–0.9	0.03–1

*Under investigation at NASA/Lewis

The ion propulsion system for a space tug can be designed so it will interface with payloads delivered to LEO by the space shuttle. The payloads would be transferred to a space tug which, over a period of 6 to 12 months, would follow a spiral trajectory to synchronous orbit.

The development of a light-weight solar collector array for the SSPS can also provide power for the ion propulsion stage. During transit, between LEO and synchronous orbit, the space plasma and Van Allen belt radiation environment must be traversed. An ion stage system with a constant-thrust spiral trajectory would spend about 100 days in the Van Allen belt. Exposure to both these environments would cause some degradation in the solar cell array power output. In the plasma, the degradation would be temporary but in the radiation belt it would be cumulative and permanent. Deployment of the ion stage to an altitude of about 300 miles obviates the effect of the space plasma environments. There are two alternatives to minimize these effects: (1) trajectory shaping to minimize the time spent in the Van Allen radiation belt, and (2) the use of auxiliary propulsion fast-trip sources (such as chemical stages). Cost projections were previously developed for a number of candidate propulsion stages applicable for transportation from Earth-to-LEO and LEO-to-geosynchronous orbit.

Preliminary flight profiles and propulsion vehicle combinations have been investigated to expose the type of environmental and gross operational factors that must be considered in evolving a space transportation system for an SSPS. It is evident that both orbital assembly considerations and alternative transportation modes are involved in the definition and development of a low-cost, high-volume, Earth-to-synchronous orbit advanced space transportation system. This definition effort will involve considerable system engineering and technology development which will greatly benefit from the activities currently under way in support of the present space shuttle development.

Orbital Assembly. — Operations in space involving assembly have been limited to docking two actively maneuvering vehicles together. However, studies of modular space stations which dock a number of similar masses to form a large complex in orbit have been performed. In general, to effect a mating, attitude control is required on both target and docking vehicles.

After a docking vehicle contacts a space assembly, several things happen: (1) the assembly experiences loads; (2) modular elements deflect relative to each other, and (3) possibly disruptive control forces and G loads are transmitted throughout the space assembly. In general, the weak link in a space assembly is the docking interface, since it has a smaller cross-sectional area than the prime construction element. Relatively large space assemblies have been analysed in modular space station studies and found to be controllable during assembly (34), providing that a prescribed build-up sequence is followed and that grossly asymmetric configurations are avoided. In addition, there is a significant interplay between docking contact velocities, target and docking vehicle flight control, and docking mechanism characteristics. In general, direct vehicle docking contact velocities of about 0.5 fps and manipulator docking contact velocities of 0.1 fps are compatible with currently planned spacecraft systems. The large, more flexible SSPS-type spacecraft will require significantly lower contact velocities because of the size and flexibility of the system and its potential damping characteristics. Zero or near-zero contact velocities may be required, together with appropriate docking or joining mechanisms and control techniques during assembly. Because a considerable portion of the SSPS structure may be part of the power distribution system, new joining and assembly techniques may be required.

Assembly sequences and modes and desirable assembly altitudes have to be identified to define the assembly requirements for the SSPS's large area and light-weight structure. Once the framework for the basic SSPS structure has been developed, including the potential sizes and sub-element characteristics of major components, such as the solar collector arrays, the transmitting antenna and structural members, alternative assembly sequences, modes, and altitudes can be evaluated. These could include automatic or operator assembly options, and assessments as to the portions of the assembly process to be carried out at low-Earth orbit altitudes, at intermediate or at synchronous altitudes. With this knowledge, the appropriate control techniques during the assembly process can then be identified and developed.

Key Environmental Issues

Resource Use. — The SSPS represents an approach to power generation which does not use a terrestrial energy source. Thus, the environmental degradation associated with mining, transportation, or refining of natural energy sources is absent. Natural resources will have to be used to produce the components for the SSPS and the propellants for transportation to orbit. Nearly all the materials to be used for the components are abundant. For each SSPS, the rare materials required, such as platinum or gallium, would be less than 2% of the supply available to the United States per year as Table 23 shows.

TABLE 23

SSPS CRITICAL MATERIALS SUPPLY

Component	Material	1b/10⁷ kW SSPS	Potential Supply to U.S. (lb/year)
Amplitron Cathode	Platinum (1 mil)	2,000	100,000
Amplitron Magnet			
	(a) Samarium	0.2×10^6	6×10^5
	(b) Samarium & Praseodymium	$0.2 \times 10^6 +$	6×10^6
	(c) Sm Pr. + Mischmetal	$0.2 \times 10^6 +$	12×10^6
Rectenna Diodes	Gallium Arsenide	1,000	5,000

+ Combinations of materials with Samarium are estimated to be heavier.

Note: Most Critical Materials Goal: 2% of Supply to U.S.

In addition, energy resource requirements for the SSPS will be minimal. Thus, the time required for one operational SSPS to pay back the energy expended during the construction phases, including raw materials, manufacturing processes, component assembly, space transportation, and ground support facilities has been estimated to be:

Propellants	6 months
Solar cells	3 months
Ground support equipment	3 months

Effects at the Receiving Antenna Site

Local ecological and environmental effects at the receiving antenna site include (a) possible hazards to organisms in the receiving area due to the microwave energy received, and (b) the effects of added heat load due to microwave-to-dc conversion inefficiencies on: both the local fauna and flora (ecological effects) and the atmosphere (urban heat island effects).

Microwave Biological Effects. -- The consideration of possible biological effects of the received microwave power beam involves:

- Limiting exposure of humans *occupationally* exposed to an acceptable level;

- Limiting *exposure of the general public* in accessible regions around the receiving site to acceptable levels for which not only direct exposure effects on the body must be considered, but also any hazards or annoyance which the microwave radiation can cause through interference (e.g., with medical devices like the pacemaker or consumer or industrial devices for which microwave radiation can cause an electro-explosive hazard);

133

- Limiting exposure in the beam to prevent disturbance to birds and aircraft. The latter would involve hazards due to incidental interference with electronic equipment or electro-explosive hazards; and

- Limiting exposure of organisms on the ground in the region of the receiver and beyond to minimize local ecological effects due to selective elimination of some classes of animals.

a. Organism Exposure

Microwave energy received at the ground could conceivably be lethal to a limited size range of organisms. Depending on the location, these could include snakes, lizards, rodents, and some insects. These organisms usually constitute the most significant members of the food chain including the major herbivores (rodents) and some important carnivores (snakes). If microwave energy at the anticipated levels does prove to be lethal to some organisms in the receiver area, then the results could be felt over a wider region. The receiver area could act as a biological sink attracting in-migration of organisms replacing those killed inside the area. Alternatively, or in addition, this area could act as a source of other organisms which might reproduce rapidly if their predators are reduced, and migrate out of the receiver region. Metal screening to attentuate microwave energy beneath the rectenna is recommended as an inexpensive precautionary measure which would prevent such effects.

b. Human Exposure

The potential hazard of human exposure to the microwave beam relates both to those persons occupied in maintaining and operating the receiving site and the space crew maintaining and operating the space transmitting station.

In the Western World, for many years the accepted exposure standard for microwaves has been $10mW/cm^2$ (i.e., the maximum power density to be measured in space accessible to workers when measured in the absence of the workers). The existence of much lower permissible exposure levels (e.g., down to $10\mu W/cm^2$) in Eastern Europe for long-duration exposure will pose a potential problem in the planning of the satellite solar power system both on the ground and in space.

Many Western scientists believe future research will verify the validity of the $10mW/cm^2$ exposure standard. Some Czech scientists believe that the $10\mu W/cm^2$ long duration standard could be safely raised. Leading Russian scientists believe in the validity of their standard.

It is likely that the general public beyond a 15-km radius will be exposed to less than $10\mu W/cm^2$. This would seem acceptable even under Soviet or Czech standards, particularly since the microwave radiation will be cw and not pulsed.

It is necessary to understand biological effects even at very low exposure levels. Programs to provide such understanding are being considered by the Office of Telecommunications Policy.

134

c. Bird Exposure

There is evidence that birds can be affected at levels of the order of 25-40 mW/cm^2 at least at X-band, with pulsed radiation. The evidence suggests an avoidance reaction by birds. Such effects could possibly be exploited to inhibit birds from flying into the receiving antenna area. On the other hand, birds may be attracted by a possible pleasant warming sensation of microwave radiation at least in some climates. More research would have to be conducted to more clearly determine effects of microwaves on birds before deciding the best choice of system parameters for minimum interference of and by birds.

d. Aircraft Passenger Exposure

The possible effects of microwave exposure on aircraft flying through the beam must be considered. Excessive body exposure to aircraft occupants is unlikely because of the shielding effect of the metal fuselage. Nevertheless, studies would be required to confirm the absence of peculiar focusing effects within the aircraft. The limited duration of flight through such a beam would support the likelihood of no serious problem with regard to human safety from radiation effects.

e. Atmospheric Attenuation

An investigation of the atmospheric attenuation of power from the main power beam was conducted to determine the attenuation in the four SSPS ground locations under investigation, to compare them with each other, and to recommend a frequency for SSPS power transmission and ground reception for a satellite orbital location at the stable node near 123° West in synchronous equatorial orbit.

The three major inputs for the investigation were as follows:

(1) The atmospheric attenuation of microwave power equations and the computations for specific conditions were taken from Reference 39* and additional data were obtained from the files of Dr. Vince Falcone, AFCRL, as well as from References 40 and 41.

(2) Rainfall rates, as reported in Reference 42, based upon the procedures of Reference 43, using as basic data Reference 44 with comparative and general information from References 43, 45, 46 and 47 and the files of Mr. Norman Sissewin and associates at AFCRL.

---

*An extension of the work in the above reference as applied to analysis of microwave attenuation is based on analytical procedures developed by Dr. Falcone of the AFCRL, Microwave Physics Branch, Bedford, Massachusetts.

(3) An approximate model of a thunderstorm, as suggested by Dr. Kenneth Hardy, Weather Radar Branch, Meteorology Laboratory, AFCRL, L.G. Hanscom Field, Rosemary Dyer, Research Physicist, and Dr. Arnold A. Barnes, Jr., Research Physicist, to be considered in formulating the atmospheric model for attenuation assessment. This model is represented by Figure 89.

Table 24 presents the data derived from input (2) and indicates specific peak rainfall rates at each of the SSPS locations:

SSPS S.W.   (Desert Southwest)
SSPS N.W.   (Northwest)
SSPS M.W.   (Midwest)
SSPS N.E.   (Northeast)

The geographic locations of a typical site in each geographical area are as described and the latitudes and longitudes are as shown in Reference 48.

TABLE 24

RAINFALL RATES

Station	Representative of SSPS Ground Station	Peak Rainfall Rates (mm/min) from input (2) for three low probabilities of occurrence		
		0.1%	0.5%	1.0%
Flagstaff Arizona		8.26 (Jul)	0.04 (Jul)	0.03 (Oct)
Phoenix, Arizona	SSPS SW	0.59 (Jul)	0.14 (Jul)	0.03 (Jul)
Tucson, Arizona		0.51 (Jul)	0.11 (Jul)	0.04 (Oct)
Walla Walla, Wash.		0.21 (Oct)	0.12 (Oct)	0.01 (Oct)
Yakima, Wash.	SSPS NW	0.15	*	*
Huntington, W. Va.	SSPS MW	0.60 (Jul)	0.20 (Jul)	0.09 (Jul)
Williamsport, Penn.	SSPS NE	0.51 (Jul)	0.16 (Jul)	0.08 (Jul)

* < 0.015 mm/min.

SUMMARY ASSUMPTIONS TO BE EMPLOYED FOR FURTHER INVESTIGATION

Name of Location	0.1% Probability		1% Probability	
	mm/min	mm/hr.	mm/min	mm/hr.
SSPS SW	60	35.	0.033	2.
SSPS NW	.20	12.	0.033	2.
SSPS MW	60	35.	0.033	2.
SSPS NE	60	35.	0.033	2.

Note: Summary assumptions are close approximations to data from nearby appropriate stations and are rounded off to coincide with available data from the analyses of Dr. Falcone.

136

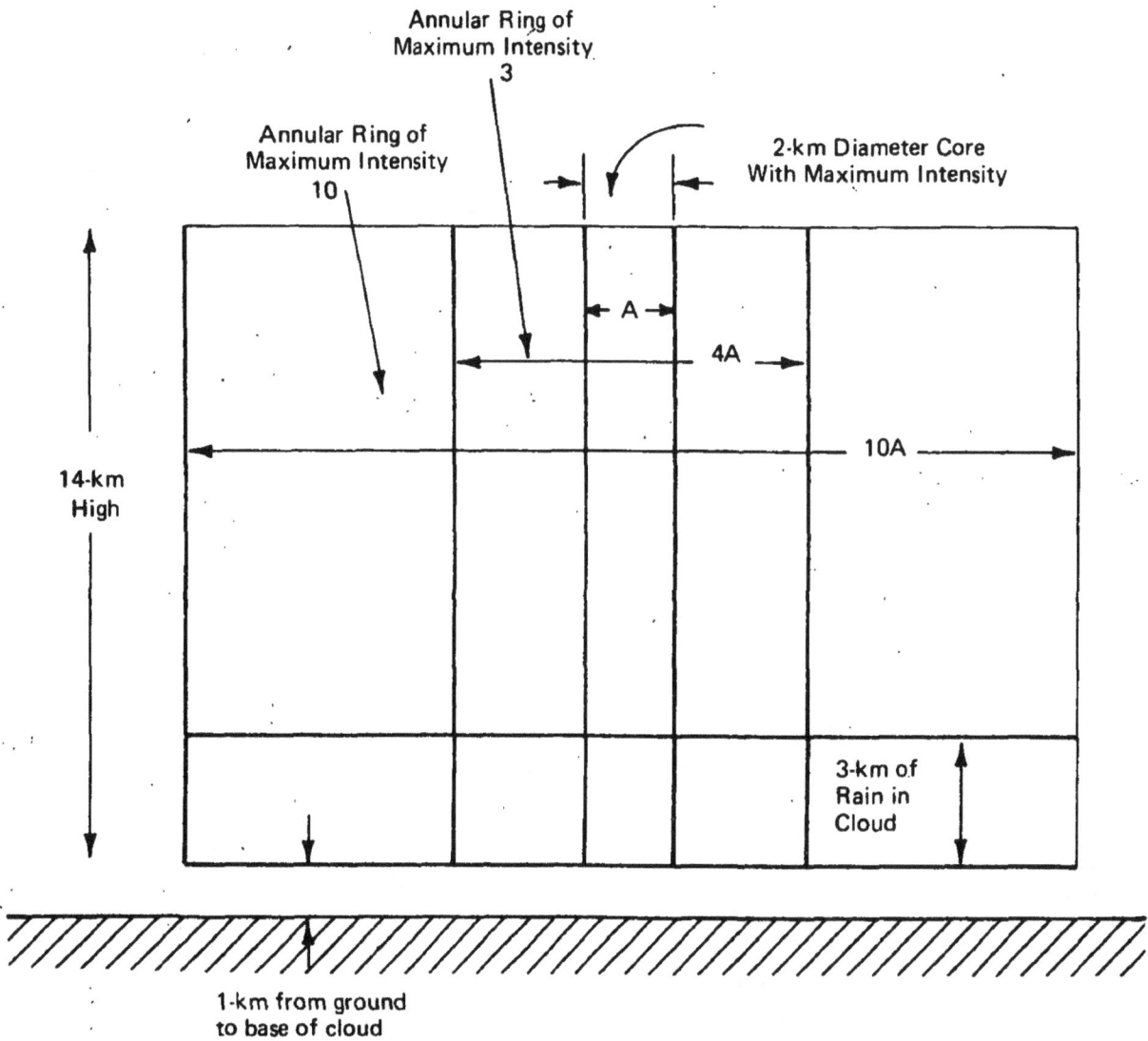

Annular Ring of
Maximum Intensity
3

Annular Ring of
Maximum Intensity
10

2-km Diameter Core
With Maximum Intensity

A

4A

10A

14-km
High

3-km of
Rain in
Cloud

1-km from ground
to base of cloud

FIGURE 89. – CRUDE MODEL OF THUNDERSTORM

137

The angles of elevation (α) above the horizon of the line of sight to the orbiting power-transmitting antenna are as follows:

Location	Name	Latitude	Longitude	Elevation Angle to Satellite α	Nadir Angle = 90−α
Southwest	SSPS S.W.	33° N	113°30'W	50°	40°
Northwest	SSPS N.W.	46° N	119°30'W	37°	53°
Midwest	SSPS M.W.	36°30'N	87°40'W	31°	59°
Northeast	SSPS N.W.	41°30'N	78°30'W	20°	70°

Figure 90 presents the data at each of the four ground base locations derived from input (1). Specific percentage attenuation is shown at each station for 3.3 GHz for comparative purposes.

There is a significant imprecision evidenced by the data points at the 2- and 3-GHz frequencies not fitting the nominal curve. This is believed to be associated primarily with the computer program's utilization of interpolated data from Table I of Reference 41, which is particularly imprecise in the 2- and 3-GHz region. The programs are currently being updated utilizing data from Reference 42.

The cloud models used in the program assumed the maximum intensities associated with the 2-km diameter core to be applied over the entire field of the microwave power beam which would, of course, result in conservative estimates of the total power attenuation.

Based on the information obtained, the following preliminary conclusions were reached:

1. There is significantly higher attenuation in the Northeast and Midwest than in the Southwest and Northwest for both high and low probability of rainfall. This is due largely to the East being generally more humid and the longer beam paths through the atmosphere.

2. The Northwest has the lowest attenuation for the low probability events. This is due primarily to the minimum precipitable water content achieved by the cooling of the air mass as it moves from West to East over the Olympic Mountains and the Cascade range and the precipitation of most of the precipitable water content on the western side of the mountains.

3. The Southwest has the lowest attenuation at the higher probabilities of occurrence due to a low average precipitation and also the shortest distance for transmission of power through the atmosphere.

The two Western locations would be favored from the atmospheric attenuation point of view. Except for the Northwest, frequencies above 4 GHz are significantly attenuated. Attenuation grows progressively worse as frequency increases. As long as the frequency is in the region of 3.3 GHz, the penalties for any ground location compared to the "optimum," although significant, are not overwhelming.

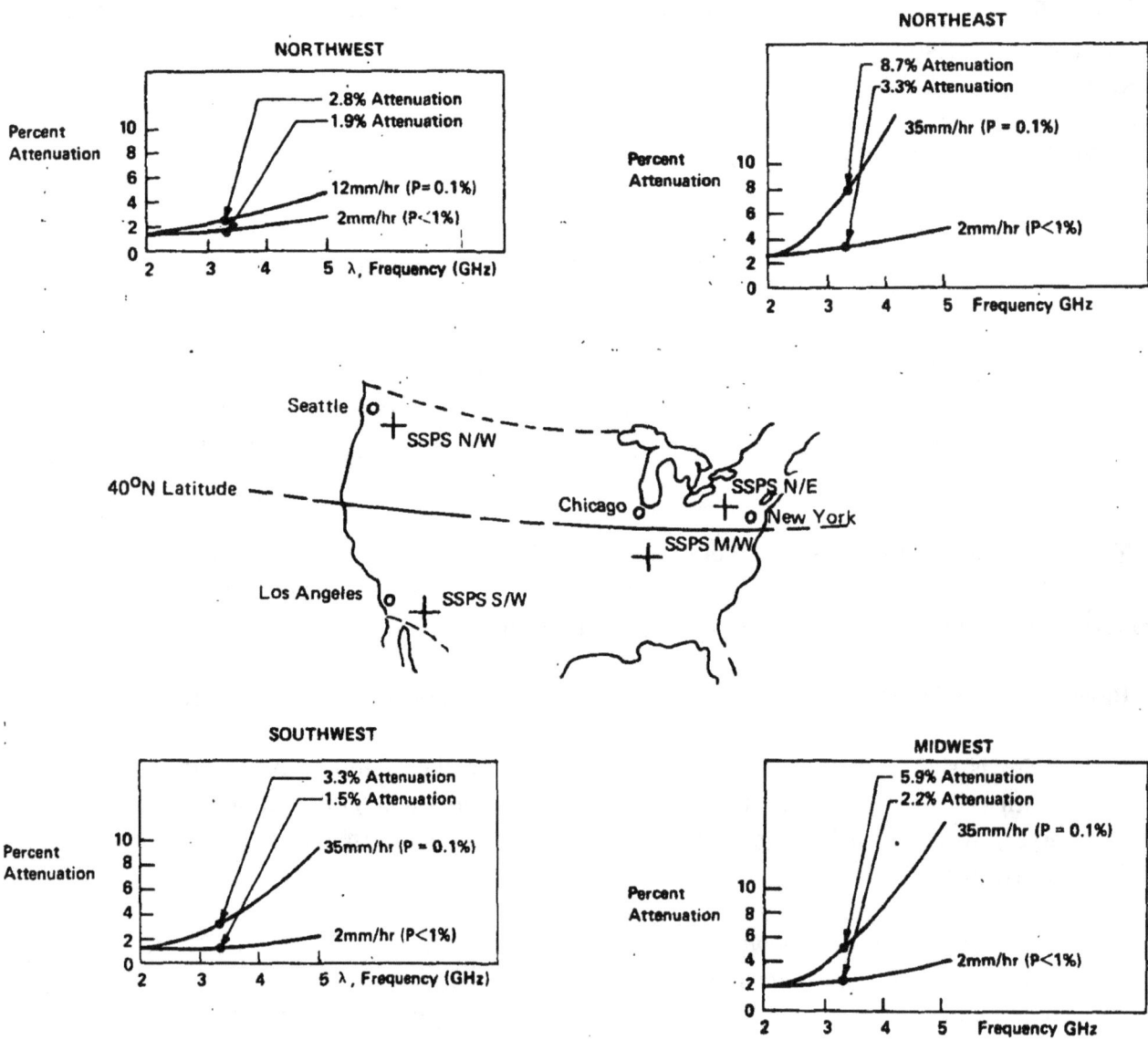

FIGURE 90. — ATMOSPHERIC ATTENUATION

139

Ecological and Environmental Effects of Added Heat. — Possible environmental and ecological effects include (1) the impact on the local flora and fauna of any significant increase in the experienced heat load caused by the rejection of waste heat at the receiving site and (2) "urban heat island" effects on the local atmosphere. Ecological effects can be evaluated by comparing energy losses from the receiving antenna with the energy environments characteristic of representative regions on Earth. To evaluate the possibility of a significant "heat island" effect, these losses can be compared with the energy consumption density for cities, on the basis that a significant heat island effect would be observed only if the energy loss rates were comparable to those experienced in cities where urban heat island effects have been suspected or observed.

We have concluded that these environmental and ecological effects of heat are negligible. Thermal energy losses from the receiving antenna are very low in absolute terms, corresponding to mean heat losses in the range of 3 to 18 W/m^2, depending on the size and efficiency of the receiver. More important, these energy losses are negligible when contrasted with the natural thermal energy flux occurring virtually everywhere on the surface of the Earth at any time of the year. Similarly, the loss is trivial when compared to the energy released by cities, and consequently there will be no urban heat island effect due to the presence of the receiver.

The assumptions made and data supporting these conclusions are presented below.

a. Receiving Antenna Heat Release

A small fraction of the transmitted microwave power is absorbed in the atmosphere, and 90% of the remainder falls within the receiving antenna on Earth, which may vary from 3.85 to 7.5 km in radius. Assuming an 85 to 90% efficiency of rf-to-dc conversion, an energy loss totaling 5.6 to 8.4 $\times 10^8$ watts will have to be dissipated at the receiving site, which corresponds to a mean heat loss in the range 3.22 and 18 W/m^2, depending on size and efficiency. Power is not evenly distributed over the receiving antenna. Depending on the characteristics of the transmitter, the peak microwave power in the center of the beam could reach 330 W/m^2, which corresponds to a peak heat loss of about 50 W/m^2, assuming the lower conversion efficiency.

The actual area of the heat transfer surfaces of the antenna has not been determined, but if the area is about 20% of the area covered by the antennas, exceedingly low heat transfer rates would be sufficient to dissipate the waste heat to the ambient environment. At this very low heat release per unit area natural air convection would obviate an environmental or ecological threat.

b. Availability of Solar Energy

The energy to be released from the antenna may be compared with the solar energy available at the surface of the Earth in latitudes from 20° to 60° N. This is equal to the solar energy available at the top of the atmosphere on a horizontal surface, less that absorbed in the atmosphere and that reflected back to space by cloud tops, and scattered by particles and other components of the atmosphere. Atmospheric losses and scattering from the atmosphere are extremely variable over the surface of the Earth, but average about 25% of the value of undepleted radiation. The latter varies seasonally with latitude. Column 1 of Table 25 gives the daily mean value of undepleted solar radiation for the winter and summer solstice as a function of latitude for the Northern Hemisphere. In higher latitudes the seasonal variations are great. Thus, at 60 degrees, the range is from 24 to 465 W/m², while at 20 degrees it varies from 280 to 420 W/m².

TABLE 25

SOLAR ENERGY

Latitude (degree)	Annual Range in Daily Mean Undepleted Solar Radiation (W/m²)	Range of Mean Annual Net Radiation (W/m²)
60	24-465	26- 52
50	82-445	40- 65
40	140-440	52-100
30	220-430	75-130
20	280-420	100-150

Receiving antenna losses: mean 3-18 W/m² max 20 W/m²

Sources: Gates (50) after Budyko (49).

Relatively little of the solar energy reaching the surface is expended in work. Some is lost as reflected solar energy, and some is reradiated as IR energy given off by the atmosphere or by the warm surface. The net radiation is the difference between solar insolation and the reflected, visible, and reradiated IR energy. It has been estimated for several locations on the earth by Budyko (49). Representative values of the annual mean net radiation are shown in column 2 of Table 26. It will be seen that these values are very large compared to the estimated energy loss from the receiving antenna. Thus, at 60 degrees the range is from 26 to about 52 W/m², while nearer the equator at 20 degrees it varies from about 100 to 150 W/m². These net radiation values are large compared to the estimated energy losses of the receiving antenna. Since it is unlikely that a large percentage of the energy lost from the antenna will, in fact, be absorbed by the ground, the effect on the net radiation at the site will be much less than the relative values of the local net radiation and the estimated heat losses imply. This is best illustrated in the next section where the total environmental energy fluxes for representative environments with the quantity of energy added by the presence of a receiver are compared.

c. Factors Determining "Environmental Temperature"

Examination of Table 25 convincingly supports the arguments that the mean antenna losses are low, but there are worst possible situations that might be encountered; e.g., winter when the solar insolation is lowest, and nighttime, when there is no sun at all. Furthermore, although net radiation and solar insolation are important factors in determining the heat load impinging on local flora and fauna, they are *not*, in fact, of primary direct importance in determining what might be called the "environmental temperature." This is due to the fact that such organisms actually intercept both incoming solar radiation from the sky *and* outgoing solar radiation that is reflected from the ground and other reflecting surfaces. In addition, thermal radiation — the incoming and

141

outgoing flux of IR energy radiated from ground, plants, clouds, and sky – is of equal and often greater importance in determining the effective heat load than is solar radiation. This fact is clarified in Table 26 where values of the IR radiation from the ground and sky are compared with relevant values of solar radiation.

The incident solar radiation can be very large for very short periods of time, as shown in Table 26A where representative maximum values of solar insolation at the surface of the Earth are shown, together with the solar constant, which is included for comparison purposes. A high mountain top near the equator may experience insolation values as high as 1325 W/m² for very short durations, and solar insolation on a sunny summer noon at sea level in the mid-latitudes may be as great as 700 W/m². These, however, are maxima, and the average insolation during the day will be much less, even in summer.

Although the contribution of solar radiation to the energy environment can be large at times, it is, in fact, usually a minor contributor to the total. The latter includes, in addition, all the IR energy radiated from warm objects in the environment – soil, rocks, structures, leaves, and shrubs – as well as IR energy from the sky. The IR energy radiated from these objects can be calculated by means of the Stephan-Boltzmann equation, since all these objects closely resemble blackbodies. Table 26B lists some representative values of the IR energy sources.

Extremely large amounts of energy are available from very cold frozen ground. Thus, at -20°C such ground radiates over 10 times the maximum estimated energy that will be lost from the center of the SSPS receiving antenna. In contrast, a warm rock in a desert can reach a temperature of 60°C, and when it does it will be radiating IR at a rate equivalent to the maximum solar insolation, or 700 W/m².

IR energy is radiated downward from the sky in a somewhat more complicated fashion. Clouds radiate as blackbodies and, consequently, the energy available from them depends on the temperature of the radiating surface, which varies with cloud type and cloud height. IR energy is also available from CO_2, water vapor, and ozone in the atmosphere, and consequently on cloudless days will depend on vapor pressure and the temperature throughout the atmosphere. The IR available from the sky has been frequently measured. The lowest radiant temperature of the Alaska winter nighttime sky observed during one series of experiments was -80°C, observed on a few cloudless nights (50). Thus the net IR loss from the ground corresponds to nearly twice the maximum energy loss in the center of the receiving antenna.

Table 27, from Gates (50), indicates the manner in which the solar and IR components of the energy environment combine to affect organisms, in this case a horizontal leaf which absorbs energy on two surfaces. The absolute energy flux to which such a leaf is subjected is given in the second column from the right. The values range from 1846 W/m² on a sand dune during a sunny day to 670 W/m² on the same sand dune on a clear night. The coolest place shown – the interior of a deciduous forest – is characterized by an absolute energy flux of more than 900 W/m² during a summer day. Although these values are for representative hypothetical environments in mid latitudes in summer, consideration of the data of Table 26 indicates that even in winter the

142

environmental temperature will, in general, be extremely high compared to the energy that will be contributed by the receiving antenna.

TABLE 26

**REPRESENTATIVE ELEMENTS OF THE RADIATION ENVIRONMENT
DETERMINING ENVIRONMENTAL TEMPERATURE**

	Energy	
	Value in Calories per Unit Area per Unit Time	(W/m^2)
A. SOLAR RADIATION		
Solar constant	$2 \; cal/cm^2 \; min$	1400
Maximum polar summer daily total of undepleted radiation	$1060 \; cal/cm^2 \; day$	498
Maximum insolation on equatoral mountain top	$1.75 \; cal/cm^2 \; min$	1325
Representative maximum summer insolation in mid-latitudes	$1 \; cal/cm^2 \; min$	700

B. THERMAL RADIATION

Ground, Leaves, etc. (Blackbody) Temperature $°C$	Energy (W/m^2)
$60°$	700
40	540
20	420
0	320
−20	235
−80	80
−100	50

C. IR RADIATION FROM NIGHTTIME SKY

Mid latitude summer, clear	304
Mid latitude summer, cloudy	374
Alaska winter radiant temperature $-80°C$ to $0°C$	80 to 320

d. Added Heat Stress in Normally Adverse Environments

It may be argued that, while the heat contributed by the antenna will have little adverse effect in environments where air temperatures are moderate and there is an adequate supply of soil moisture, additions of energy to environments where organisms already experience significant heat stress and conditions are already marginal may cause noticeable damage. Thus, it may be argued that to locate the receiver in a desert may be to impose an insuperable burden on a region which is ecologically fragile under the best circumstances.

The possible impact of the maximum added heat stress on the local water balance can be illustrated as follows. One mechanism by which plants reject heat is evapo-transpiration. During

TABLE 27

RADIATION REGIMES FOR HYPOTHETICAL ENVIRONMENTS

Conditions	Air Temperature (°C)	Ground Temperature (°C)	Albedo (%)	Vapor Pressure (mb)	Energy Flux (W/m²)						Absorption by Horizontal Leaf (2 surfaces)
					Solar Sky	Reflected Solar	Downward Long Wave	Upward Long Wave	Net Flux	Absolute Flux	
Sand Dunes:											
Clear Day	30	40	30	36.1	700	-209	388	552	326	1846	1673
Cloudy Day	30	30	30	36.1	397	-118	464	485	258	1465	1356
Clear Night	15	10	—	15.2	—	—	303	368	66	670	650
Cloudy Night	15	15	—	15.2	—	—	372	388	16	760	737
Open Clearing in Forest with Grass											
Clear Day	30	40	7	36.1	700	-49	388	552	484	1685	1540
Bog with Lake Clear Day:											
Over water	30	25	9	36.1	700	-63	388	403	574	1595	1456
Over mat	30	40	2	36.1	700	-14	388	552	519	1651	1510
Interior of Deciduous Forest	temperature of tree crown										
Clear Day	23	20	—	—	35	0	444	423	56	902	869
Clear Day	35	30	—	—	35	0	514	486	29	1034	997

Source: Adapted from Gates (50)

144

photosynthesis, stomata on the surface of leaves are open to permit absorption of CO_2 and release of O_2. As long as the stomata are open, there will be a continuous loss of water vapor to the air, the quantity depending on the temperature and relative humidity of the air and the number of open stomata. If the soil water supply declines and insufficient water is available to satisfy the demand, the stomata close and photosynthesis ceases. At this point the temperature of the leaf may rise, since heat is no longer being lost by evaporation. If the heat being absorbed from the environment cannot be rejected as IR radiation and/or as sensible heat conducted to the air, the leaves may gradually wilt and die.

In extremely hot dry climates where the heat load is too intense part of the time due to lack of sufficient ground water, the growth of many plants may be limited to periods of rainfall, so that annual droughts are survived in the seed stage. Other plants are adapted to storing water for long periods of time, and still others can survive surprisingly long periods of drying out.

The effect of the receiving antenna's heat loss can be partly evaluated by considering the additional pressure on the water balance which would result if the energy absorbed by the plants were rejected by them as latent heat, that is, in terms of the extra water which would be needed for this purpose. About one half the total energy lost by the receiver would be absorbed by the plants on the ground below the receiving array. This would represent a heat rejection problem only during the daytime. The result, therefore, would be equivalent to increasing the potential evapo-transpiration by an amount equivalent to 25% of the heat rejected. The corresponding values are 10 to 65 mm per year for the range of mean energy loss values and about 170 mm for the assumed maximum of 50 w/m² in the center of the beam.

The effect on typical desert communities of this increase in potential evapo-transpiration may be judged by comparing it with the factors determining the water balance in these communities. The factors of interest are:

(a) the potential evapo-transpiration, as calculated from the rainfall (humidity) and temperature;

(b) the actual evapo-transpiration, which depends on the potential evapo-transpiration and the availability of soil water (and which can be roughly calculated knowing the rainfall and soil type);

(c) the evapo-transpiration deficit, which indicates the capacity of the vegetation community to reject heat by other means than evaporation, or to survive high temperatures by any of a number of adaptive mechanisms.

During some months precipitation may be much higher than necessary to meet the needs of the community (for example, in winter), so a surplus in water may result even when potential evapo-transpiration exceeds actual evapo-transpiration for the year.

145

If typical desert communities show ranges in the evapo-transpiration deficit which exceed the estimated water stress imposed by the receiving antenna, then locating a receiving antenna in such a community is not likely to disrupt it excessively. The result may be a slight change in the composition of the local community in the direction of similar communities in slightly drier climes. Such a shift would probably be observable only to an expert, if at all.

Table 28 shows that this is the case. The water budget components for representative vegetation in the western United States are tabulated. The ranges of these components for communities in the northeastern (colder) and southwestern (dryer and hotter) regions are shown. The evapo-transpiration deficit for the year is shown in the second column from the right. Deficits characteristic of these desert communities vary from 142 mm of water for a sagebrush community in Utah to 1328 mm of water for an alkali sink community in California. The range of deficits observed for each type of community is large, varying from about 140 for the sagebrush to about 350 mm for one type of creosote bush community found in parts of California and Nevada. Since these total deficits and ranges of deficits are large compared to the estimated 10 to 65 mm load imposed by the receiving antenna mean heat losses, we may conclude that the effects on such communities will be small. They will resemble the natural effects due to local variations in topography, and be almost undiscernible.

e. Environmental Effects of Local Heat Losses

The primary environmental effects of heat losses are modifications to local climate caused by changes in the sensible or latent heat transferred to the atmosphere which, in turn, cause the air to rise, and/or, change its moisture regime. Such heat may also cause modifications in rainfall patterns in the local environment. Some effects attributed to cities may also be due to the emission of condensation nuclei and to changes in the surface roughness, which also affect turbulent motions of the air. These have been reviewed by Landsberg (52). Although many studies have indicated generally drier and hotter air near cities, other effects such as modifications of rainfall patterns are more difficult to discern reliably, and have not been extensively studied. (See, for example, Huff and Chagnon (53).

The possibly deleterious effects of receiving antenna heat losses can be indicated by comparing them to the heat losses characteristic of cities in which urban heat island effects have been observed or suspected.

The latter may be determined by examining the total energy consumption (in terms of fuel use) of selected cities or large, densely populated megalapoli as a function of area. Table 29 lists the relevant factors for several regions of the globe, together with the mean annual net radiation at those points. This table shows that many regions reach or exceed the estimated maximum receiving antenna heat loss of 50 W/m^2, including Manhattan, (with 630 W/m^2) and Moscow (with 127 W/m^2). We may conclude from this table that the receiving antenna by itself will have no heat island effect, although sites of energy use may.

146

TABLE 28

RANGE IN COMPONENTS OF THE WATER BALANCE FOR DESERT AND DRY GRASSLAND COMMUNITIES IN WESTERN UNITED STATES[†]

Vegetation Type	Location	Altitude (m)	Ppt (mm)	Avg Temp (°C)	PE (mm)	AE (mm)	Deficit (mm)	Surplus
Shadscale	Colo-Utah	1246-1570	120-255 (192)	6.7-11.4 (10.5)	595-744 (711)	120-255 (192)	340-624 (519)	0
Shadscale	Nevada	1210-1332	98-136 (114)	10.1-12.1 (11.0)	551-728 (649)	98-136 114	439-576 (535)	0
Sagebrush	Utah-Wyoming	1341-2186	230-442 (373)	2.0-11.3 (7.0)	385-693 (571)	230-360 (313)	142-380 (259)	0-111 (66)
Sagebrush	E. Washington	259-747	187-372 (270)	7.9-11.2 (9.9)	600-737 (678)	187-359 (252)	320-510 (426)	0-57 (17)
Creosote Bush	SE Calif-Nev	-6-658	78-104 96	17.7-22.9 (20.7)	933-1275 (1127)	78-104 (96)	829-1191 (1030)	0
Alkali Sink	SE Calif	-14-245	53-129	17.4-24.9	973-1381	53-129	844-1328	0

†Means shown in parentheses. PE is potential evapo-transpiration; AE is actual evapo-transpiration.

Source: Major (51)

147

TABLE 29

ENERGY CONSUMPTION (EC) DENSITY IN SELECTED INDUSTRIAL AND URBAN AREAS

	Area (km²)	Population 10⁶	EC Density (W/m²)	EC per Capita, E_h	Average Net Radiation (W/m²)
Nordrhein-Westfalen	34,039	16.84	4.2	8.0	50
Same, industrial area only	10,296	11.27	10.2	8.9^b	51
West Berlin	234	2.3	21.3	2.0	57
Moscow	878	6.42	127	16.8^b	42
Sheffield (1952)	48	0.5	19	1.6	46
Hamburg	747	1.83	12.6^a	5.0	55
Cincinnati	200^a	0.54	26	9.3	99
Los Angeles County	10,000	7.0	7.5	10.3	108
Los Angeles	3,500^a	7.0	21	10.3	108
New York, Manhattan	59	1.7	630	21.0	93
21 metropolitan areas (Washington-Boston)	87,000	33	4.4	11.2^c	~90
Fairbanks, Alaska	37	0.03	18.5	21.8	18

a. Building area only. b. Related to industrial production. c. Eastern United States.

Source: MIT Press (54)

Land Usage in the Receiving Antenna. – The receiving antenna could be compatible with a wide range of land uses, from farming to high-intensity residential, commercial, and industrial uses. This should be kept in mind when decisions regarding its location have to be made. Thus, no matter where the receiving antenna is located, there will be a requirement to transmit its power to the complexes where it is to be used. The cost of the land and of the transmission system will be major factors; low-return land usages, such as farming, will do little to offset their investments, whereas high-intensity uses – for example, in a city center – will offer a large return as well as provide savings vis-à-vis the transmission systems.

Present trends in living patterns favor very high-density residential use combined with some industry and considerable commerce. These are ecologically sound in that they are well adapted to mass transit and preserve land for farming and open space. Serious consideration should be given to combining the receiving antenna with the city it is meant to serve.

The concept of a single span covering over large cities has been advanced by the inventor of the geodesic dome, Buckminster Fuller, and is presently undergoing experiments. Fuller claims that the geodesic concept results in economies of scale in the sense that the strength-to-weight ratio of large domes increases with size rather than the opposite. Roofed cities have distinct advantages over conventional ones; for example, there would be complete weather protection resulting in savings in snow removal, less work time lost, fewer vehicular accidents, etc. Fuller claims, for example, that the costs of a 2-mile diameter dome over Manhattan could be recovered in 10 years from the savings in snow removal costs alone. There would also be large savings in energy consumed for heating and cooling because of the drastically reduced surface area of the city.

If the receiving antenna were to be integrated into a roof structure over a city, several advantages and possibilities appear: no new land would be consumed by the rectenna nor would long-distance transmission lines be necessary, with the city receiving its power from the satellite directly.

Tidal marshes and similar coastal wetlands have been suggested as possible sites for a receiving antenna, on the grounds that these are "waste" of little economic value. However, the extremely great ecological importance of such sites effectively prohibits this or any other use of them. The existence of offshore fisheries is wholly dependent on continued unimpaired functioning of such coastal wetlands. Unlike a meadow or a field of crops, these wetlands could be severely disturbed by the disruption associated with construction of a receiving antenna, which would require deep footings, and might even entail dredging. Even minor amounts of travel for example, through, a salt marsh for the purpose of routine maintainence or inspection could have a serious cumulative negative effect. Furthermore, the presence of the antenna structures and the associated RF screening would certainly reduce the light intensity on the ground; reductions of up to 20% are possible, and these would result in corresponding — and intolerable — reductions in overall productivity.

Open land under the antenna could be used for a variety of compatible agricultural purposes. The lowering of light intensity would generally have an adverse effect on the agricultural value of the land, although there are some crops which do not thrive in full sunlight (for example, shade-grown tobacco). For other kinds of crops, a lowered overall productivity might be tolerated.

Stratospheric Pollution with Shuttle Vehicle Exhaust Products. —

a. Introduction

Considerable attention has recently been paid to the potentially harmful effects of supersonic transport (SST) exhausts in the stratosphere, although current opinion seems to hold that these may have been overestimated at first.

The potential hazards of SST operation which have received the greatest attention include:

(1) The injection of particulate matter or gases which combine to produce particulate matter which may either reduce solar insolation by increasing the population of scattering particles, or increase the temperature of the atmosphere by increasing the population of particles which absorb solar radiation. Comparisons of the maximum possible effects of SST's with the effects of volcanic explosions have shown the former to be very small compared to the latter (54).

(2) The injection of water vapor and NO_x which are involved in the complex sequence of chemical reactions governing the abundance of ozone in the region from 20 to 35 km where ozone is most abundant. Increases in either constituent are believed to result in lowering of the mean abundance of ozone, but there is great uncertainty regarding the roles of each and little agreement between critics. The ozone abundance is a very important factor in determining the amount and wavelength of potential damaging UV radiation which reaches the surface of the earth.

The possible effects of NO and H_2O injected into the stratosphere by the shuttle vehicles can be identified by comparing their emissions with those estimated for the SST fleets. The actual effects of any given rate of injection of either component are difficult to determine because of (1) uncertainties regarding the vertical and horizontal movements in the stratosphere which govern the rate at which the injected material is distributed within the stratosphere, and ultimately removed from it, (2) lack of reliable experimental observations of the composition of the stratosphere as a function of altitude, season, and location of the surface of the globe, and (3) great uncertainties regarding the nature of the chemical and photochemical reactions which determine the abundances of chemical species involved in the ozone equilibrium.

Vertical mixing in the stratosphere is very slow and declines with increasing altitude. Consequently, gases injected into the stratosphere will accumulate, and even a low annual rate of injection will yield a large equilibrium value at very high altitudes. One way of evaluating the effects of injection of constituents is to employ available knowledge of the mean lifetime at the altitude in question to compute an equilibrium value and compare it with estimates of the "natural" abundance of the gas in question. This approach is limited by the uncertainties regarding the composition of the stratosphere. It can be used for the main engine water vapor exhaust, but uncertainties regarding the abundance of NO at any level in the stratosphere and of the importance of NO in the ozone equilibrium suggest that other criteria should be employed for this constituent.

b. Water Vapor Pollution

The main engines of the launch vehicle burn for 8 minutes and consume 1.6×10^6 pounds of fuel in the form of hydrogen and oxygen which combine to produce a like quantity of water. After 30 seconds, the vehicle reaches an altitude of 5,000 feet and in another 30 seconds it reaches the tropopause, which is assumed for the purposes of this calculation to be 10 km. During the remaining 7 minutes it climbs an additional 100 km. 1.4×10^5 pounds of H_2O are assumed to be emitted in each 10 km depth of the atmosphere above the tropopause. 360 launches are assumed per year. Residence times in the stratosphere are not known with certainty: Martell (55) estimates residence time to be one month at the tropopause, about one to two years at the 20 km level, and 4 to 20 years at 50 km (the stratopause). Table 30 shows the calculated increase of water vapor to be expected in each 10-km slice of the stratosphere, assuming the range of residence times listed. These are compared with the best available estimates of the natural abundance of water vapor at these altitudes, namely 2×10^{-6} kg/kg of air up to 30 km and 5×10^{-6} kg/kg of air above this level. The water vapor increment due to main engine shuttle vehicle exhaust is seen to be insignificant in the altitude region where ozone is most abundant. Thus, the increment corresponds to between 0.02

150

and 0.1% in the altitude range of 30 to 40 km and less at lower altitudes. Most of the ozone in the atmosphere is located between about 23 and 35 km. These water vapor injections are unlikely to have any direct effect on the ozone content as a whole. However, they will be concentrated in an area above the launch site, and the consequences of this localization should be explored.

TABLE 30

SHUTTLE VEHICLE WATER VAPOR INJECTION INTO THE STRATOSPHERE

Altitude	Assumed Residence Time (years)	Mass of Atmosphere (10^{18} gm)	Natural H_2O Contents (10^{12} gm)	Added H_2O (at Equilibrium) (10^{12} gm)	Percent Increment
10-20 km	0.1- 1	1000	2000	0.0024 to 0.024	1.2 to 12 x 10^{-5}
20-30	2- 4	200	400	0.042 to 0.090	0.01 to 0.02
30-40	2-10	50	250	0.042 to 0.29	0.02 to 0.1
40-50	4-20	1	5	0.090 to 0.46	1.8 to 10

The percentage increment in water vapor is large in the highest region of the stratosphere near 50 km, due to the fact that the atmosphere is extremely thin and consequently its total water content is low, even though the mixing ratio is about twice that observed at lower levels.

The chemistry of water vapor in the upper stratosphere has been studied but there is great uncertainty regarding the possible consequences of increments in water vapor on the order of 10%. Water vapor is photodissociated to form hydrogen, hydroxyl, and hydroperoxyl radicals and hydrogen peroxide molecules which will react with ozone, and molecular and atomic oxygen. The latter constituent is abundant at this level. Since some of the NO_x at lower levels is produced in the mesosphere and carried downward through the region in question, it is conceivable that changes in the water vapor content will influence the natural flux of NO_x to the level of the ozone layer. Consequently, the effects of shuttle flight water vapor injection in the region of 40 to 50 km should receive further study.

Booster engines using solid fuel are also employed during the first two minutes after launch. These consume a total of 3×10^{-6} pounds of fuel per launch, of which 4×10^5 pounds consist of CO, HCl, and NO. We have been unable to obtain information regarding the chemical form of the remaining exhaust products. Even if all of it is water, it is unlikely that there will be any adverse effect on the stratosphere due to water vapor pollution.

According to Martell (55), vertical eddy mixing is very rapid throughout the mesosphere (50 to 80 km) and the lower thermosphere (80 to 100 km) and for this reason the effect of water vapor emissions in these regions has not been calculated. The region of the atmosphere from the stratopause to 100 km in altitude contains only 0.1% of the total mass of the atmosphere, or about 5×18^{18} grams. The annual injection of water vapor into this region will be about 12×10^9 gms.

Injection of water and other exhaust products will, of course, be very localized, so that large increases in the abundance of water vapor will be expected in the very small region near the launch

151

point. The possible local effects have not been considered, since there is little reliable information regarding the rate at which they will be horizontally mixed.

c. NO Pollution of the Stratosphere

Booster engines will emit approximately 40 metric tons of NO into the stratosphere in the region from 10 to 24 km per launch, or a total of 2.4×10^{11} grams per year. These emissions are on the order of those to be expected from a 500-unit SST fleet, and are in approximately the same altitude region. Some authors have argued that this is a modest increase when compared with the probable "natural" concentrations of NO at these levels. However, there are few direct measurements of NO or NO_2 in the stratosphere, and all of the estimates of abundance are derived from models which are recognized as being simplifications of the real situation. The strongest critic of the SST program is Harold Johnston (56, 57), who argues that the significance of the SST NO injection should be determined by comparing it with the natural flux of NO from the upper atmosphere, where it is produced by photolytic decomposition of N_2O. Ozone is less abundant in the region from 10 to about 35 km than calculations of the equilibrium concentration that would occur in a stagnant atmosphere if there were no sinks due to the presence of water, methane, and oxides of nitrogen. Johnston argues that this decrease of ozone is due to the NO carried down from the stratopause. Others suggest that some of the discrepancies must be due to mixing in the stratosphere or to other effects.

The best (e.g., most conservative) approach to evaluating the effects of booster engine NO pollution in the stratosphere is to employ Johnston's arguments and see whether the shuttle vehicle NO emissions are large compared to the natural fluxes he and others have calculated. Table 31 shows these comparisons. Several "SST fleets" have been employed in the literature. The earliest controversy concerned a 500-unit fleet of SST's of the U.S. type, which consume 66 tons of fuel per hour while cruising in the stratosphere and emit from 13.2 (latest estimate) to 42 lb of NO per 1000 pounds of fuel. The Concorde is reported as consuming 33,000 (SCEP (58)) to 18,000 pounds of fuel per hour, and to emit 13.2 to 15 pounds of NO per hour. All SST's are assumed to average 7 hours per day in the stratosphere at an altitude of about 20 km.

Table 31 shows that the NO emissions from the booster are large and within the range of SST emissions that have caused concern.

Thus NO emissions may constitute a pollution problem in the stratosphere if Johnston's approach is the correct one. Consequently, the SST controversy should be followed closely, and the localization of these emissions should be evaluated if possible.

Other Booster Emissions. —The remaining 2.6×10^6 pounds of pollutants emitted by the booster should be identified and their effects on the stratosphere and troposhere should be evaluated.

TABLE 31

NATURAL AND ARTIFICIAL FLUX OF NO INTO THE STRATOSPHERE

Source	Flux (10^6 metric tons/year)
Natural[a]	0.2 to 2.0
SST Fleets[b]	
U.S. Type[c]	1.2 to 3.6
Concorde[d]	0.16 to 0.4
Shuttle Boosters (360 launches/year)	1.4

---

a. From Johnson (57), Crutzen (59), Nicolet and Vergison (60), and McElroy and McConnell (61).

b. Each SST is assumed to cruise at approximately 20 km for 7 hours per day.

c. U.S. type SST's consume 66 tons of fuel per hour and emit 13 to 42 pounds of NO per 1000 pounds of fuel.

d. Concordes consume 9 to 16 tons of fuel per hour and emit 13 to 15 pounds of NO per 1000 pounds of fuel.

Tropospheric Pollution. — The effects of HCl, NO, or CO in the lower troposphere in the regions of the launch site have not been considered. During the first 30 seconds after launch when the vehicle travels from the ground to an altitude of about 5000 feet, it emits about 138 tons of CO, 126 tons of HCl, and 18 tons of NO. Although these quantities are low compared to other large sources of these components, we feel that somewhat closer analysis of their effects is warranted, and that other constituents of the exhaust should be considered.

Key Economic Issues

Key Cost Considerations. — Capital cost projections for a system as complex as an SSPS must at this time be recognized as preliminary because of the limited detailed engineering effort that has gone into the design of the components and subsystems. It is possible to assign cost ranges to the most significant parts of the system which include silicon solar cells, microwave generation, transmission and rectification, and the space transportation. The following section reviews the basis for capital cost projections for specific elements of the SSPS.

a. Solar Energy Conversion

The present cost of silicon solar cells for use in spacecraft — about \$175/W — is prohibitive. New methods for producing single-crystal silicon and mass-production assembly techniques will have to be developed to reach the goal of less than \$1/W. Based on the experience of present

153

manufacturers, there is a high probability that low-cost, high-volume silicon single crystals of the desired thickness can be produced once there is a large enough market (62). Hand assembly techniques of solar cells, which are adequate to meet the present very small demand, will have to be replaced by automated methods similar to those that have been perfected for the production of other semiconductor devices.

Since the preliminary cost projections for the SSPS solar photovoltaic array (63) were prepared in November 1971, there have been several developments which have led to new inputs to solar cell cost estimates. These new cost estimates will have an impact on the solar array costs that were used in the overall system cost analysis. In addition to these new cost projections, there have been design changes influencing the concentration ratios and leading to a decrease in the bussing requirements. Therefore, a re-evaluation has been made of the cost projections to point out areas and to update these projections.

Cell efficiency, or power output, for a 50-μm (2-mil) thick cell was projected to be 19.7% and have a power output of 26.7 mW/cm^2 (Figure 3.2-7, Ref. 63) at beginning of life and at 300°K. This is assumed to still be an accurate projection.

Since a 50-μm thick cell is still being assumed, there is no change in the cell weight which was listed as 16.8 mg/cm^2 (Figure 3.3-12, Ref. 63).

The method of growing crystals of silicon and obtaining thin 50-μm thick material was assumed to be either hex rod or pulled web (Table 3.4-2, Ref. 63). Recent studies with edge-defined, film-fed growth (EFG) of silicon ribbons has provided new information for the SSPS cost analysis (64, 65). A comparison was made of the Reference 64 cost analysis of the ribbon, the cells fabricated from the ribbon and the SSPS report. The raw EFG ribbon cost was projected to be 0.25 cent per cm^2, while the pulled web cost (Table 3.4-2, Ref. 63) was 0.20 cent per cm^2. This shows very close agreement to the lowest costs projected, although the hex rod material costs were much higher. Therefore, it appears that the hex rod approach is too conservative and the ribbon cost of 0.25 cent per cm^2 is more representative as a silicon crystal raw material cost. The cell fabrication cost used in Reference 63 was 1.0 cent per cm^2 and, when added to the web material cost of 0.2 cent per cm^2, amounted to a total solar cell cost of 1.2 cent per cm^2 (Table 3.4-2, Ref. 63). The EFG ribbon cell cost analysis indicated that the total cell cost is estimated as 0.38 cent per cm^2. This is a significantly lower cost and will be used in the updated analysis.

The complete blanket cost was obtained (Table 3.4-2, Ref. 63) by adding 0.3 cent per cm^2 for laminating the cover and printed circuit substrate to the total cell cost of 1.2 cent per cm^2, thus resulting in a blanket cost of 1.5 cent per cm^2. For the updated analysis, the lamination cost will remain at 0.3 cent per cm^2, and this is added to the 0.38 cent per cm^2 cost to total 0.68 cent per cm^2 for the complete solar cell array blanket. Using the power output number of 26.7 mW/cm^2 (STC), this results in a blanket cost to power ratio of $255 per kilowatt compared to the $566 per kilowatt figure used in the preliminary SSPS cost analysis.

The next consideration concerns the effects of these new blanket cost projections on the solar collector costs. The baseline concentrator structure is of a flat-plate channel design with a concentration ratio of N = 2.06. At this concentration ratio, the power (delivered in space after 5 years of life) from the blanket corresponds to 38 W/cm² (Figure 3.5-20, Ref. 63) or an area-to-power ratio of 26.3 cm² per watt (Figure 3.5-23, Ref. 63).

The reflector or concentrator surface area required to achieve a concentration ratio of N = 2.06 for the new channel design is 2.12 times the blanket area. Therefore, multiplying the new blanket area per watt figure of 26.3 cm² per watt by 2.12 results in a reflector area per watt figure of 55.7 cm² per watt (generated in space after 5 years).

Based on the analysis of the effects of cell layout and interconnection pattern on bussing weight, a substantial decrease in the array bus weight and, consequently costs, can be projected. The layout utilizing the blanket as the bus for the complete length of the array results in a bus weight of 4×10^7 grams for the 8×10^9 watt (in space) solar collector array or a weight-power ratio of 5×10^{-3} gm/W.

Using the updated costs, the solar collector costs can be summarized and compared to the original SSPS cost projections (Table 3.8-1, Ref. 63). Since these new cost projections are based on ribbon cell technology and a flat-plate, channel-type concentrating mirror design, the cost comparisons were made with the web-type solar cells using the low-cost projections for such cells and the N = 3 petal-type concentrating mirrors. These cost comparisons are shown in Table 32.

The cost projection of 0.352 cent per watt of power generated on Earth is substantially lower than the 0.672 cost per watt figure used in the 1971 projection. Therefore, the cost projection of $352 per kilowatt is assumed for the solar collector array cost.

Cost projections based on actual experience in the production of single-crystal silicon solar cells are presented in Figure 91. The 2x2 cm cells are widely used on unmanned spacecraft. The 2x6 cm cells were produced for the Apollo telescope mount portion of the "Skylab" spacecraft. The 2-inch diameter cell represents the costs of solar cells for terrestrial applications produced from 2-inch diameter wafers of single-crystal silicon boules.

This perspective on experience for solar cells can be used to test the reasonableness of the solar cell cost projections. The 1971 low cost projection corresponding to 1.5 cents per cm² fits on a 75% slope line while the 1973 cost projection of 0.7 cents per cm² fits on a 72% slope line. Both of these cost projections are reasonable slopes, since experience indicates that typical cost reduction curves follow about a 70% slope. Therefore, this test of reasonableness for the solar cell costs indicates that these projections follow past industry experience (66).

The concept that solar arrays can be manufactured inexpensively in the future was emphasized by Paul A. Berman (67): ". . .it seems almost inconceivable that such a simple thing as a solar array substrate with printed circuit interconnections and wiring upon which cells are mounted in some

TABLE 32

COST PROJECTION COMPARISONS FOR TWO SSPS SOLAR COLLECTOR ARRAY CONFIGURATIONS

NEW 1973 N = 2.0 Channel-Type Concentrating Mirror Design (Baseline)
OLD 1971 N = 3.0 Petal-Type Concentrating Mirror Design *

System	Area (cm²)/W		gm/W		$/W	
	1973 (N=2)	1971 (N=3)	1973 (N=2)	1971 (N=3)	1973 (N=2)	1971 (N=3)
Blanket	26.3	21.0	0.742	0.592	0.179	0.315
Reflector	55.7	84.0	0.111	0.168	0.036	0.054
Bus and Support	—	—	0.030	0.120	0.005	0.079
Subtotals (power generated in orbit)			0.883 g/W	0.880 g/W	0.220/W	0.448/W
Orbit to Earth Factor (equivalent power generated on Earth)			(1.6X)	(1.5X)	(1.6X)	(1.5X)
Totals			1.410 g/W	1.320 g/W	0.352/W	0.672/W

*Ref. 63

156

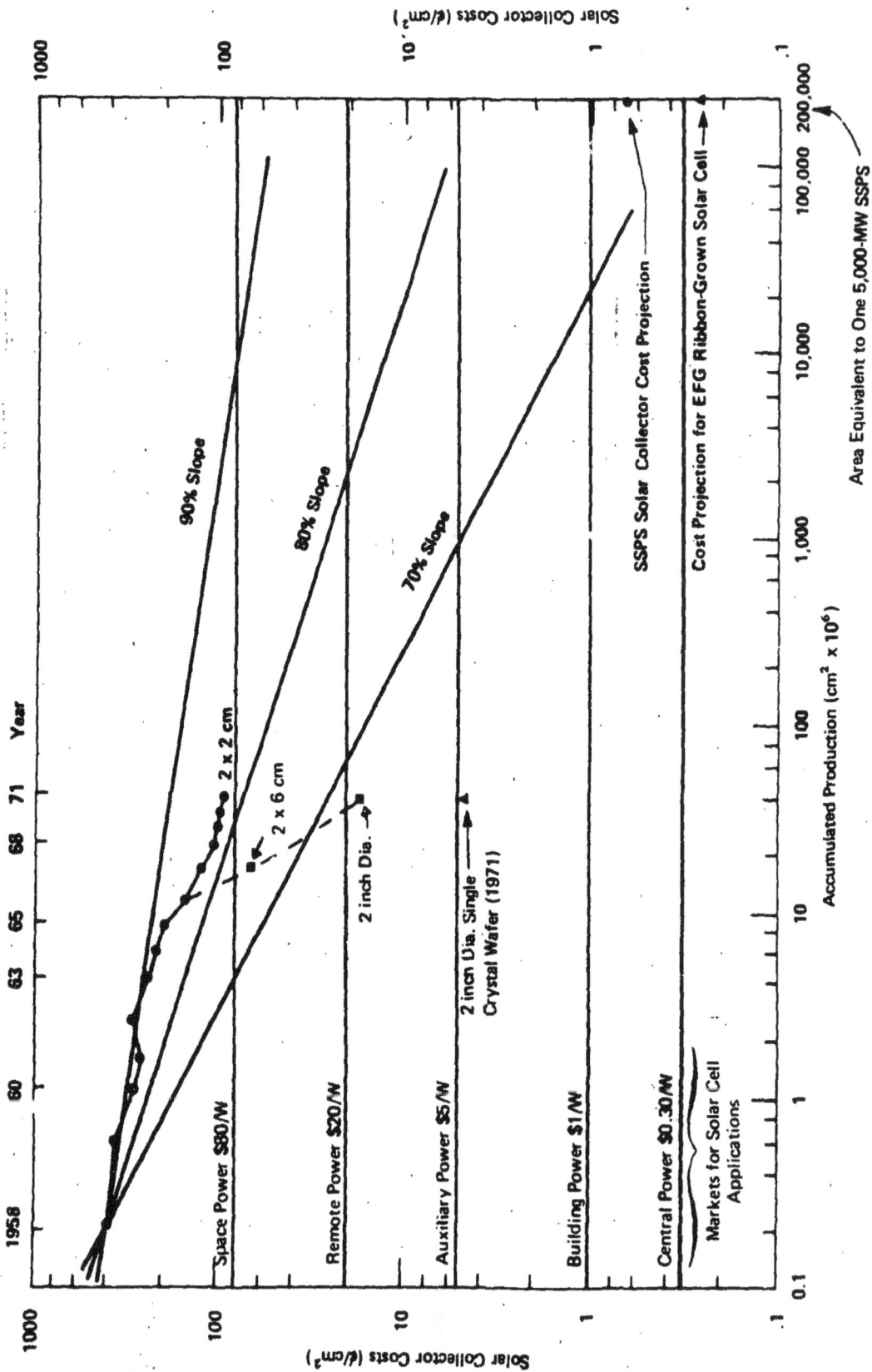

FIGURE 91. — SILICON SOLAR CELL COST PROJECTIONS FOR SSPS SOLAR COLLECTOR ARRAY

157

simple, economical manner, and over which some inexpensive protective layer is positioned, having no moving parts and using no exotic materials, cannot be made for a few dollars a square meter rather than the thousands of dollars per square meter experienced in the space program. . ."

b. Space Transportation System

Space transportation costs projected for an operational SSPS system are based upon the extensive studies of advanced transportation systems, conducted by NASA and Grumman, prior to the selection of the current space shuttle system. In addition, the results of technology development programs related to advanced propulsion techniques (e.g., ion/electric propulsion) were used to expand the base of potential transportation options that might be considered for an SSPS program. The foundation for transportation cost projections for an operational SSPS are therefore based upon an extensive advanced transportation system "data bank" that has been developed over the past few years. We are therefore confident that the cost projections represent realistic estimates of technically-achievable goals in the 1985-1990 time period.

Three basic phases lead to an operational SSPS program; they are:

- Technology Development/Verification,
- Prototype System Development, and
- Operational System Development.

As regards space transportation aspects related to these phases, it has been assumed that ground-to-low Earth orbit (LEO) transport of payloads would be accomplished by the space shuttle system presently under development by NASA *for the technology development and prototype phases only.* For the operational SSPS system, it is assumed that a *new* low-operating cost transportation system would be developed, compatible with the high traffic volume needs. It is expected that this new transportation system would be an evolutionary growth of the present space shuttle, and that its development costs would be amortized over the quantity of operational SSPS platforms expected to be operating in space in the future.

Projected delivery costs for an operational SSPS were developed by examining several alternative combinations of space transportation elements for delivery of payloads to synchronous orbit where final assembly of the SSPS was assumed to take place. The transportation elements examined included all-chemical and mixed chemical-ion transportation systems:

(1) The fully reuseable (booster and orbiter) two-stage Earth orbital shuttle studied by NASA during 1971 and 1972, generally referred to as the "Phase B Shuttle." The operational costs of payload delivery to LEO of this system are less than half those of the partially reuseable shuttle currently under development by NASA.

(2) The chemical propulsion stage (CPS), a large space-based reuseable "space tug" that launches itself into orbit after staging from the reuseable booster.

(3) The Space Tug (ST), a small space-based reuseable stage that would be a second stage to the CPS. It would be delivered to orbit by the EOS or the CPS.

(4) The ion propulsion stage (IPS), a large space-based solar electric ion stage that would deliver SSPS subassemblies from LEO to synchronous orbit. Projected cost and performance data for the IPS were obtained from NASA-Lewis and industry sources; the overall specific power system weight (lb/kW) of the IPS was based on SSPS solar array technology.

The all-chemical STS consisting of the EOS, CPS, and ST yielded delivery costs to synchronous orbit between $300 and $500 per pound. The mixed chemical-ion STS consisting of the EOS, ST and IPS projected costs of $100/lb and became the baseline transportation system for operational SSPS delivery. This baseline system thus projects a transportation cost of $2.5 billion for delivery of a 5000-MW operational SSPS.

A comparison of the capabilities of the space shuttle now under development with the requirements for SSPS shuttles is shown in the table on page 132.

The space shuttle now under development provides the necessary first step towards a high-volume, low-cost transportation system for an operational SSPS. Further, advanced single-stage Earth orbit shuttle concepts have recently been proposed (68, 69) that show promise of reducing projected SSPS transportation costs to less than $100/lb. These large reductions in operational costs will provide the incentive to invest in the development in a new space transportation system, when the decision to deploy an operational system is to be made in the early 1990's.

Recent assessments of SSPS assembly operations have indicated that a desirable location for SSPS assembly is a medium altitude orbit (5000 to 7000 n. mi.) located above the high radiation regions that would affect solar cell performance. In this concept the SSPS flies itself to synchronous orbit from the medium altitude (assembly) orbit using its own ion propulsion system.

The combination of advanced single stage to orbit transportation system elements and assembly operations at less than synchronous altitudes provide optimism that a $50/lb goal, for transportation costs associated with the delivery of the operational SSPS, is potentially achievable by the 1990's.

Based on these considerations, the transportation to orbit and assembly cost projections are as follows:

(1) Prototype SSPS

   (a) $1380/kW for SSPS baseline design and "current" space shuttle transportation costs and projected assembly costs, and

159

(b) $950/kW for SSPS baseline design, including projected weight saving improvements and "current" space shuttle transportation costs and projected assembly costs.

(2) Operational SSPS

$250/kW for SSPS baseline design, including projected weight saving improvements, an advanced space transportation system and projected assembly costs.

c. Microwave Generation, Transmission and Rectification Costs

Cost estimates for the microwave generators can be based on the substantial experience with state-of-the-art microwave devices. The cost estimates on the transmitting antenna are less firm, because detailed design of the antenna has not yet been accomplished. Space structures have a high cost when measured in terms of dollars per pound of structure compared to the cost of commercially produced structures. However, most of the structure of the transmitting antenna will be highly repetitive and will be produced by mass production techniques. Because of the need for light-weight structures, the largest part of the cost will be in light-weight, high-performance materials. Table 33 provides cost estimates on the microwave generators and the transmitting antenna, indicating the range of costs based on the information available at the present.

Table 34 provides the statistics on the receiving antenna on which the cost projections shown in Table 35 are based. The cost projections for the receiving antenna have been derived on the basis of material and labor costs for elements which can be produced by highly automated methods. As an example of the potential for automation, the number of diodes in one SSPS receiving antenna would require a year's production at the rate of manufacture of 18,000 diodes per minute. Yet the amount of material required would be less than 4,000 pounds because of the very small chip size in the diode. In addition, the structure is simple in design (expanded metal mesh) on which the very light-weight rectifier elements are mounted.

d. Capital and Operating Costs

The technical reports cited in References 2 through 10 describe the rationale for the baseline design for the SSPS. Workable versions of each component for the SSPS exist today, or can be built, although some will entail considerable development. The costs of the major components, such as the solar cells, microwave generators, and rectifiers, can be drastically reduced. Programs are already under way to achieve cost reductions. Further reductions can be achieved by mass production, as demonstrated by the successful evolution from development to production for a wide range of products (71).

The estimates for the capital costs are shown in Table 36. The three columns describe an appraisal of the possible cost spread. A figure in the 0.25 column reflects the belief that there is one chance in four that the cost would be equal to or less than the figure cited. Similarly, the entries in the 0.50 and 0.75 columns suggest that there are two and three chances, respectively, out of four that the cost will be at least as favorable as the figure cited.

TABLE 33

COST ESTIMATES ON TRANSMITTING ANTENNA

Item	Basis of Cost	RF Cost Projection $/kW of RF Output		
		Low	Medium	High
Tube				
RF circuit	Production experience on 1-kW electronic oven tube applied to 5-kW Amplitron without vacuum envelope	3	4	6
Magnetic circuit	$90.00/lb for Sm-Co finished parts and design in section 4. 3. 3. 5. of Ref. 9	3	6	7
Pyrolytic graphite radiators	$100.00/lb for finished parts and design in section 4. 3. 3. 5. of Ref. 9	7	11	15
Self-regulation	Educated guess	2	6	8
Control Electronics Projection		10	20	30
Balance of Structure				
High weight of 0.343 kg/kW	$200.00/kg and Table 2 (Ref. 9)			69
Target weight of 0.176 kg/kW	$300.00/kg and Table 2 (Ref. 9)		53	
Low weight of 0.069 kg/kW	$500.00/kg and Table 2 (Ref. 9)	35		
Contingency		10	30	45
Total		$70	$130	$180

Source: Ref. 9

161

TABLE 34

RECEIVING ANTENNA STATISTICS

Wavelength	10 cm
Rectenna diameter	7.4 km
Total area	43×10^6 m^2
Average power density	
10,000 MW dc	232 W/m^2
5,000 MW dc	116 W/m^2
Total number of elements	1.23×10^{10}
Antenna element density	284/m^2
Maximum power per element	
10,000 MW dc	3.0 W
5,000 MW dc	1.5 W

Source: Reference 70.

TABLE 35

COST ELEMENTS OF RECEIVING ANTENNA

Per Square Meter

Schottky-barrier diodes	$ 2.84/m^2
Rectenna circuit and assembly of diodes into circuit	3.16/m^2
Supporting structure and final assembly	5.50/m^2
Total	$ 11.50/m^2

Per Kilowatts of DC Output

10,000 MW (232 W/m^2)	$ 50.00/kW
5,000 MW (116 W/m^2)	100.00/kW

Note: The cost is a function of area at the power density levels of interest.

Source: Reference 70.

TABLE 36

ESTIMATED CAPITAL COSTS
($/kW)

	p = .25	p = .50	p = .75
Solar Array	310	500	700
Transmitting Antenna (including microwave generators)	80	130	175
Receiving Antenna	50	100	150
Transportation to Orbit	190	450	810
Assembly	60	100	140
Composite Total	900	1300	1800

Source: SSPS Team Estimates

These cost estimates were developed against the baseline design of 5,000 MW. The extreme modularity of the solar cell arrays and microwave transmission system is such that the unit costs (i.e., dollars per kilowatt) may be insensitive to the power output of the SSPS.

If the assumption is made that the prototype system is represented by 0.75 column and the operational system by the 0.25 column, it would appear that the costs are comparable to the prototype and operational costs of other competing energy-production systems.

The composite figure is more than a simple sum. The figure for 0.25 which represents the overall cost of the system must be greater than the sum of individual figures in the 0.25 column — how much greater depends on the degree of independence of the individual events and the nature of the individual distribution (i.e., the curve between and beyond the points 0.25, 0.50, and 0.75). The composite figures* are based on very simple assumptions:

(1) The events are completely independent; and
(2) The distribution is nearly rectangular.

The assumption is made that the table describes the principal capital components, and that additional items will be minor or insignificant relative to the cost figures presented.

Although the details have not been worked out, the assumption can be made that the development cost of the transportation delivery system would be about $10 billion, and that the development costs for all other SSPS systems would not exceed these costs. These estimates would have to be made on a stand-alone basis; that is, no cost-sharing with other programs. The proper share of these development costs would be written off against a number of SSPS's. On the

---

*The details are described in Reference 72.

163

assumption that ultimately 100 SSPS's would be built, the burden would be $40 per kilowatt (less than 5% of the capital cost), with the entire development cost of the transportation and delivery system and the SSPS development cost charged to the program.

The SSPS development program can be planned as an option with clearly identified development and decision points. Therefore, no commitment to the commercial SSPS program need be made until the technology development, verification, and prototype development phases have been successfully concluded.

Key Economic Considerations. – The three key economic issues that should be addressed when making an economic comparison of the SSPS with other means of generating power are:

1. The costs and benefits associated with the project;
2. The macro-economic interindustry effect produced by the project; and
3. The consumption effects created by the project.

The costs and benefits associated with the project are evaluated to determine the economic feasibility of an investment of this type. Such an evaluation asks the question whether the expenditure of funds on this project will produce a product which will earn the total economy a satisfactory rate of return and thereby be considered as a worthy undertaking. This analysis must take into account the social and environmental side effects of both the old and the new technology.

The macro-economic interindustry effects produced by the project are examined to analyze the effects such an investment might have on the structure of the economy as a whole. This analysis can best be accomplished by using an input-output model of the economy to test the industry effects.

The consumption effects created by the project will be reflected in both the cost/benefit analysis and the analysis of macro-economic interindustry effects, but because of its importance, it is an issue worthy of separate consideration and attention. It is concerned with total energy requirements, the impact on the capital, management, and labor resources of the U.S. economy, and with the demand for raw material and the effect on prices of those materials in the face of a massive demand.

These key issues relate not only to an efficiency criterion, where an optimal allocation of scarce resources is the sole objective, but also to indicate the effects associated with large-scale changes that might be created in the economic structure. At the very least an analysis of industry and consumption effects would identify groups that would be affected.

a. Cost/Benefit Analysis

The first part of the economic evaluation of the SSPS consists of a cost/benefit approach to determine the value of an investment of this type. A cost/benefit approach to determine the economic feasibility of a project is very similar to an evaluation undertaken by a private corporation to determine the expected returns from a potential investment. The major distinction is that a cost/benefit analysis considers the entire macro-economy as a single unit. It asks the question whether the expenditure of funds on this project will produce a product which will earn the total economy a satisfactory rate of return and thereby be considered as a worthy undertaking.

A major feature of a macro-economic cost/benefit analysis which distinguishes it from investment evaluation on a micro level is the quantitative treatment of external effects, or more precisely, a social benefit or cost that is an outgrowth of a particular investment in which the values of the benefit or cost cannot be captured within the price of the product. This feature has important implications in the evaluation of the SSPS. The presence of externalities in this case is likely to help considerably in justifying funding for development of this option for power generation, even if capital costs are comparable to the capital costs of other power-generating methods.

Some obvious external effects which should be treated are intermediate technological benefits, the effect on international trade and the balance of payments, environmental impacts, use of non-renewable resources, and energy pay-back. Such external effects are illustrative of the results of a technology assessment which would have to be carried in support of this project.

Cost/Benefit analysis has been a reliable analytical tool for policy-makers in the process of making research developmental and operational investment decisions. At the same time, its limitations are well known, perhaps best to those who have relied on it most. This does not suggest that the use of cost/benefit analysis on decisions of a long-term nature transcending national importance should be discouraged, but only that sober judgment of its promise in dealing with such decisions be applied.

Many societal problems have proven not to be amenable to vigorous quantitative benefit/cost analysis. There are too many qualitative, intangible, incommensurable values involved. The best that cost/benefit analysis can do in these kinds of circumstances is to set qualitative goals or levels of desired achievement by value judgment, and then to seek the best mix of quantifiable developmental measures consistent with and subject to these goals. So it is with environmental objectives and resource planning. The intrinsic value of a stretch of wild river, for example, can be honored as a social benefit at the constrained cost of an irrigated crop or hydroelectric power production.

The major categories of issues that should be addressed in the cost/benefit analysis are:

- The size and expected yield of the research and development investments that are required to develop an SSPS;

- The formulation of a standard of comparison which would include the cost of a composite of other means of providing power as this composite would appear in the 1990 to 2000 time frame, against which a composite system, including an SSPS, would be measured; and

165

- The development of a composite cost of power generated by a hybrid system in this time frame that included an SSPS.

1)   The Research and Development Investment Required

In the first category, the research and development investment required, the issues that have to be addressed are:

- The R&D investment required to develop an SSPS, including a portion of the development cost of interrelated programs; and

- The expected yield of research and development projects of this magnitude in order to support expectations of success.

The order of magnitude of the investment required to do the research and development for the SSPS will be comparable to the actual costs of developing nuclear power or any other large undertaking with a similar potentially significant impact on a national and eventually worldwide basis.

These cost estimates will have to include the cost of tooling and manufacturing facilities, the space transportation system as well as the cost of development efforts needed for certain components, to apply particular technologies or to further develop certain concepts. The apportioned development costs of interrelated programs that are mutually dependent upon one another for their overall justification will also have to be included.

In assessing the risk associated with a development project of this magnitude, it is important to differentiate from the total budget that portion which is truly "risk money." In the case of the overall development budget, it is conceivable that the risk portion is about 10% of that total with the remainder of the expenditures being for items such as tooling and manufacturing facilities and other hardware which could be provided on a time-phasing commensurate with increasing confidence in the positive outcome of the project. A breakdown of the total R&D budget in these terms will give greater insight into the level of risk that is being incurred.

Another means by which the level of risk can be assessed is to look to other research and development projects that were of similar magnitude to gauge their success rate in order to support the expectations of success in the SSPS project.

2) The Standard of Comparison

SSPS should be evaluated as part of a total power generation and distribution system. This is necessary for two reasons. First, preliminary cost estimates for the SSPS indicate that they are extremely capital-intensive and that the cost of power generation in mills per kilowatt-hour is almost directly coupled to the capital charges on the plant investment. Since capital charges are calculated on the basis of load factor, it is important that the SSPS plant be evaluated as part of a power supply system.

166

Secondly, systems evaluations of supply increments have replaced individual project evaluations. Up until the late 1940's the established procedure for coming to an investment decision was to consider a "least cost" engineering solution on an individual site or station basis. More recently, with the technological advances in higher voltage, longer distance transmission with the increasing entry of public systems into virtually all supply sectors, projects have become interconnected systems and intersystem coordination has demanded attention in the interests of economic efficiency and reliability. Hence, evaluations of power supply increments per se are neither useful nor relevant for planning; rather, evaluations must be made of existing systems with and without the new supply increments. This holds for conventional systems as well as mixtures of new and old technology.

A first step will be to select one or more forecasts of future demand for power as made by the Federal Power Commission, the Atomic Energy Commission, and the Edison Electric Institute. Next, it will be necessary to describe new central station power requirements by size for the entire period of forecast. Finally, the load duration curve should be examined to establish capacity versus load factor tables for any new plants.

Next, a composite cost structure in the 1990 to 2000 time frame without an SSPS, given the range of options available at that time, should be forecast. These costs should be developed for each element of the system: generation, transmission and distribution.

The costs of conventional power systems are well documented, and information such as that published by the Federal Power Commission and the Edison Electric Institute is a primary source.

All power systems consist of generation, transmission, and distribution elements and the balance design requires that tradeoffs be made, particularly between plant size and transmission costs. Recent trends toward larger and larger plants have occurred, because the larger plants can gain efficiency and economy sufficient to offset the greater transmission costs that removal from load center implies. This trend is further accelerated by advancements in transmission technology, but these alone do not suffice to raise the size of central power station plants. Recent trends in environmental standards for transmission systems may raise, rather than lower, transmission costs and thus make central station power plants located nearer to the consumer of interest indicating that plant siting flexibility, ease of energy resource supply and waste product disposal will be important considerations. The hypothesis of likely system configurations in the 1990-2000 time frame must take into consideration such cost trends. It may be necessary to consider several assumptions about transmission costs in order to see how these might affect the system structure at that time.

The composite cost structure of any power generation system should include the expense of mitigating environmental damage. It will be necessary to identify and characterize principal environmental and social issues surrounding the generation, conversion, and use of energy and assign costs to them. In some cases it will not be possible to assign rigorous values to certain extreme events, but for the more likely and usual hazards one can and should develop various damage surrogate figures, for example, insurance premiums.

Three principal kinds of environmental impacts should be considered:

- Those due to waste product disposal (e.g., combustion product emissions, radioactive materials, and heat);

- Those due to fuel mining and transport (e.g., land despoilation and fuel spills); and

- Those due to the building and deployment of apparatus (e.g., energy use, land use, and aesthetic aspects of energy generation and distribution apparatus).

These three areas should be explored with respect to conventional and proposed methods of power generation. As appropriate, quantitative factors should be stated and an assessment should be made of the gravity of each effect. More importantly, the methods proposed, or in use, to mitigate these effects should be considered and their influence on the future environmental impacts projected. Where possible, the likely cost increment for the use of advanced abatement technology should be estimated. It is recognized that most power systems have non-internalized cost components whose evaluation should be part of the decision about energy. One should be careful, however, to identify the portions of the costs that are partly internalized so that a fair and believable reckoning can be made.

While the current concern for the environment and the growth in demand for energy will be hard to reconcile, one thing is certain; viz., lessening of environmental impacts will play a greater and greater part in the selection and management of energy sources and energy-producing technology. These problems lead in three directions for relief: (1) ways to lessen the impact of present and proven energy technology on the environment; (2) palatable public policy which might lessen the demand for energy; and (3) new and more favorable technology.

The balance between old and new technology should not, and will not, be struck by the classic methods of power plant engineering and cost analysis, but must take into account the social and environmental side effects of each. While all these effects may not be understood, or even perceived, the need for decisions on central station power plant technology requires that such effects be defined and assessed. Technology cannot be selected solely because it has some aspects which are more desirable than its competitors, or because it appears to meet a particular near-term need. There is the matter of broadly defined costs and these are often barely assessed. Everything possible must be done to understand all the costs and all the major side effects of each technology.

3) Capital and Operating Costs for an SSPS as Part of a Hybrid System

The analysis of an SSPS should be based on the assumption that a major portion of the nation's electrical energy needs are to be so supplied. Long-run financing and production costs have to be estimated as well as the development costs needed to reach operational status. The costs should be annualized by consistent and conventional means and should be reflected into a mills-per-kilowatt-hour figure for a common basis of comparison. Load factor assumptions should be stated and rationalized against a forecast of power demand with typical load-duration curves.

The competitive analysis will be more variable since the amount of information known and available about any power generation differs from scheme to scheme. This difference in information is characteristic of elements in any portfolio and is part of the reason why the methodology of evaluation should be based on a portfolio concept. For example, the direct cost of proven and available technology will be readily available, whereas the possible cost of a fusion system may be essentially unknown.

Each element or system in a portfolio will have its own profile of costs, benefits, risks, and uncertainties. These have to be shown as they are appraised by knowledgeable people. At this time there are few general rules for portfolio evaluation because the utility of such general rules depends on the data available. The factors that should be dealt with, however, are (1) estimates of annualized and mills per kilowatt-hour at direct costs, development dollars to date and risk and uncertainties, and major environmental and social pro's and con's.

For the purposes of the evaluation, it will be necessary to establish the constraints and standards of performance and reliability for the SSPS component of the total power generation and distribution system. Since SSPS will be part of a hybrid system, its strengths and standards of performance will be determined by the system as a whole and therefore will be largely developed by implication for SSPS.

To compare feasible hybrid energy systems with substantial SSPS components to conventional and advanced systems of equivalent capability, a question that will have to be considered is what is the appropriate size for a single SSPS plant, given the reliability standards of the entire system. Power systems are complicated. The great Northeast power failure of 1965 made that clear to all. While some comparisons can be made between prime movers or between transmission and distribution techniques, such analysis must finally be done within a coherent systematic framework and must account for all of the activities and apparatus needed to provide the power to the consumer.

As in the case of the conventional systems, it will also be necessary for SSPS to identify and characterize the principal environmental, social, and political issues surrounding the generation, conversion, and use of energy and assign costs to them.

4) Utility Cost Accounting – Basis for Cost Comparison

Alternative generating plants can be compared in one of two ways.

- The "mills per kilowatt-hour" method that pro-rates the annual fixed costs of owning plant and equipment across the amount of energy produced in a year, and adding to this figure the direct (or variable) cost of energy generation – fuel.

- A present-value analysis that capitalizes expected future operating costs (i.e., the direct or variable costs) and adds these, as a lump sum, to the original capital outlay for the facilities.

169

The two approaches are really two sides of the same coin. In both it is necessary to make assumptions about the utilization or load factor of the plant — how many kilowatt-hours can be generated in a year for each kilowatt of installed capacity. Both require assumptions about plant lifetime, although in either method and at the discount rates typically employed, there is relatively little difference in outcome among assumptions on life that vary from 30 to 40 to 50 years.

Most utilities evaluate new generating facilities on the basis of mills per kilowatt-hour. However, they also are sensitive to, and think in terms of, "dollars per kilowatt" for the initial plant. Before nuclear plants were a viable alternative, the "dollars per kilowatt" parameter would just about by itself rank alternative fossil plant offerings; the fuel costs for all of them would be about the same. Even in the early days of nuclear plants, the fuel costs (thought of in terms of cents per million Btu's) were somewhat comparable or could be translated into rules of thumb that would say a nuclear plant could justify an additional $50-100 per kilowatt in initial cost. While dollars per kilowatt* gave a quick shorthand way of appraising the relative merits of different plants and also gave insight into the amount of capital that would have to be raised to finance new plant construction, the final comparison was usually made on the basis of mills per kilowatt-hour (or, alternatively but less frequently, on a present value calculation) that considered all the factors.

New technologies such as SSPS, fusion reactors, and breeder reactors, may have near-zero operating costs. Thus the simple rule-of-thumb "dollars per kilowatt" becomes less significant, and it is increasingly desirable to think about mills per kilowatt-hour.

Virtually all of the AEC analyses of different reactor technologies (there are perhaps about a dozen — including the high-temperature gas reactor, molten salt, and various types of breeder other than liquid metal) have been based on mills-per-killowatt-hour calculations. The numbers presented below are generally consistent with the analyses of the AEC. Typical fixed charge rates have run 13-15% per year.

5) The Role of the Load Factor

Most reactor evaluations are done on the basis of an 80% annual load factor — actual kilowatt-hours delivered per year are 80% of theoretical capacity. Eighty percent is achievable with light-water reactors which must be down for several weeks each year for refueling; indeed, it lets a reactor otherwise run at 100% capacity be "down" 10.4 weeks per year. More elaborate analyses** use a sliding scale of availability that more closely reflects utilities' experience with fossil fuel plants

---

*The dollars-per-kilowatt figure includes payments to equipment vendors, payments for design and construction services (and escalation experienced during construction), and interest during construction. In essence, the utility looks at the cost it would hypothetically pay to a third party who had borrowed money to make the necessary progress payments and had thereby incurred interest charges as one of his costs of doing business. These are the dollars per kilowatt that the public utility commissions "allow" in the rate base. This practice equalizes the differences among utilities, some of which have their own design and construction operations and most of which finance construction themselves, using new bonds and stock financing.

**For instance, the widely circulated "Report on Economic Analysis for Oyster Creek Nuclear Generating Station" issued by Jersey Central Power and Light Company in late 1963. This elaborate, detailed analysis used load factors of 88% for the first 15 years, 83% for the next five years, 67% for the next five years, and 56% for the next five years — a weighted average load factor of 78.3%.

170

(where the improved efficiency of larger units coming on-stream after the 1950's displaced earlier, less efficient units).

Load factors for use in economic comparisons have heretofore been selected primarily on the basis of technical aspects of plant life, and on the experience with a mix of generating sources — those that are more clearly "base load" (generally, higher capital costs but lower incremental costs — those units that are preferentially loaded); intermediate units (usually, older stations in semi-retirement); and "peaking units" such as gas turbines.

In the long run, however, it is necessary that the load-duration characteristics of demand be taken into consideration. Only 40% of the kilowatt capacity installed in a system can be operated in the 80-85% range; for the rest, the load simply is not there.

6) Assumptions Concerning Capital Charges

The estimated capital charge rates shown in Table 37 are based on the following assumptions:

- Financing 50% by debt, 50% by common stock equity.* This debt equity ratio is fairly typical of the industry as a whole — although in some cases the percentage of debt rises as high as 60% or 65% and some economists argue for the higher percentage.

- After-tax return on equity and the interest rates for AA bonds are based on Moody's public utilities stock and bond averages for the week of August 27, 1971 (reported in Volume 43, No. 14, August 31, 1971).

- Estimates for state and local taxes, as well as property insurance and interim replacements, are based on estimates used by the AEC and are, again, fairly typical of the industry (although property insurance has recently risen appreciably).

The resulting 16% capital charge is somewhat higher than the industry has experienced in the recent past when interest rates were lower and price-earnings ratios were higher. AEC estimates in 1967 totaled 13.7% — but this was, of course, before the serious escalation in interest rates. ADL has recently completed analyses for the Electric Research Council using 15% as more representative of conditions over the next 10 years.

---

*This ratio is based on the *market value* of the utility's common stock rather than its book value. Market value, based on the number of shares outstanding times the current market price, more nearly reflects the financing conditions of utilities as they grow in the future, rather than book value which often reflects a long series of complex accounting entries.

TABLE 37

ANNUAL EXPENSES ASSOCIATED WITH $100 OF PLANT EQUIPMENT

Expenses Deductible from Income	$/Year
State and Local Taxes	2.45
Property Insurance	0.40
Interim Replacements	0.35
Interest (8.06% on $50)	4.03
Subtotal	7.23
After-Tax Earnings to Yield an 11.4 Price/Earnings Ratio on Common Stock Sold for $50.00	4.39
Federal Income Tax (at 50% of Pretax Earnings)	4.39
Total	16.01

Source: Arthur D. Little, Inc., estimates

7) Computation Example

The calculation of the mills per kilowatt-hour is shown below. Using a 14% charge and 7000 hours per year — both figures that have been reasonably close to reality — one arrives at the easily applied *rule-of-thumb that each $100 per kilowatt of capital costs results in 2 mills per kilowatt-hour.*

$$\frac{Mills}{kwhr} = \frac{Capital\ Cost\ (\$) \times Annual\ Charges\ (\$)/\$\ of\ Capital\ Cost \times 1000\ Mills/\$}{8760\ kwhr/year \times Load\ Factors\ (fraction)}$$

For each $100 capital cost, with 80% load factor, and at 14% annual expense

$$\frac{Mills}{kwhr} = \frac{100 \times (14 \times 0.01) \times 1000}{8760 \times 0.8} = 2$$

b. Consumption of Resources

Man's total need for energy is expanding at such a rate that serious economic, social, and environmental issues are being created. It has been forecast that between now and 2001 the United States will consume more energy than it has in its entire history, and that by 2001 the annual U.S. demand for energy in all forms will double and the annual world-wide demand will probably triple. These projected increases will tax man's ability to discover, extract, and refine fuels in the huge volumes necessary to ship them safely and to dispose of waste products with minimum harm to himself and his environment. Therefore it is imperative that choices of technology be made, considering the total energy balance. Thus for any energy production method the period required to

172

pay back the energy consumed during all phases of the construction process has to be accounted for to assure that the power produced substantially exceeds the resources consumed during construction and the energy debit required to replace them.

For example, the propellants required to place SSPS in orbit or the production of solar cells are an energy tax against the system and will have to be accounted against system performance. In addition, the total energy system required for the building and operation of SSPS will have to be structured to identify and quantify the energy sources and links within the system. Preliminary estimates indicate that this energy tax against the SSPS can be met by less than one year of operation.

In addition to the necessity of viewing and analyzing SSPS in the context of the total energy balance and assessing its net impact, the net effect of SSPS on the economy will have to be determined. The demand for human and material resources will be significant and it will be necessary to establish the magnitudes of these demands and to predict the economic consequences of a decision to implement an operational SSPS system.

Certain assumptions will have to be made as to manufacturing costs and processes, for example, the degree of automation. These assumptions need to be translated into the demand for capital, management, and labor inputs. Consideration must also be given to likely industry structures, e.g., whether the solar cell production facility has to be a monopoly or whether it should be carried out by industrial enterprises. The capital requirements for these enterprises must also be estimated whenever they are not built in to the development budget.

The demand for raw materials required for the building of the SSPS system, and in particular, those materials which might be rare or exotic must be forecast. The present estimates indicate that any material required for an SSPS will not exceed 2% of the annual supply available to the United States. An estimate should also be made of what might happen to the prices of these materials in the face of increased demand.

The fundamental raw materials — hydrogen, oxygen, aluminum and silicon — are among the most abundant although there are materials, such as platinum, samarium, and cobalt which will be required in limited quantities. Clearly the supply of these materials is not likely to run out; however, there may be rarer materials required such as gallium for diodes which is today available in limited quantities. The prices for certain of these items will not necessarily rise in the face of a substantial requirement. In fact, the increases in demand for certain materials may lead to a drastic reduction in price due to changes in production processes that may be required to meet the higher demands.

By carrying the system installation projections one step further, it will be possible to calculate the raw materials requirements of the system elements. These requirements should be readily available, and, in many instances, the choice of materials will have a major effect on performance. Substantial published information exists on known and forecast reserves of raw materials which will form the basis of comparison. Price trends can be estimated by examining analogous cases for raw

173

materials and by determining how production processes for less available materials might be improved.

c. Interindustry Effects

An analysis of the effects an investment might have on the structure of the economy can best be accomplished by using an input-output (I/O) model of the economy to test the industry effect. Basically, this model answers the following question: What will the effects on other industries be? To illustrate how this question can be answered the automobile industry can serve as an example. The I/O model specifies the production technology of the automobile industry. It will show that automobiles require certain inputs from the steel industry, the plastics industry, the rubber industry and so much labor. These amounts are specified. Therefore, an increase in demand for cars will create a derived demand for products from other industries. However, these other industries also require inputs for which the demand will now be increased. The final result of the model specifies the increase in output of all the other industries which are caused by the purchase of additional cars. Table 38 gives a hypothetical I/O model of the economy.

TABLE 38

I/O MODEL OF ECONOMY

Purchasing Sectors

		X	Y	Z	Auto	Gov
Selling	X	a_1	a_2	a_3	a_4	—
	Y	a_5	a_6	a_7	a_8	—
Sectors	Z	a_9	a_{10}	a_{11}	a_{12}	—
	Auto	a_{13}	a_{14}	a_{15}	a_{16}	1

The values a_1, a_2...a_{16} are the technical coefficents of the matrix. The model shows that the auto industry sells a_{13} to industry X, a_{14} to industry Y, a_{15} to itself. The government sector represents final demand from which all other demands are derived. These coefficients actually represent the production technology. In order for the auto industry to produce one unit for the government, it must *purchase* a_4 units for industry X, a_8 units for industry Y, a_{12} units for industry Z, and a_{16} units for itself.

This model also shows how the industrial sectors interact with one another. The increase in final demand for autos will increase demand for the output of other industries, which, if they are to product more, will need more autos. This linear system can be solved mathematically and the results will specify the outputs of various industries for a given level of final demand.

A similar analysis can be used to calculate the interindustry effects of a change in the source of power. The model can now specify how much power is sold from each of the energy-producing

174

industries to each other industry, and the production requirements of each energy-producing sector — be it conventional or non-conventional. If the inputs into the production of SSPS are known, they can be placed within the model. It will then be possible to determine the interindustry changes that would result.

Below are given three hypothetical interindustry matrices with varying degrees of conversion to solar energy in SSPS.

No Conversion

	W	X	Y	Z	Coal	Solar Energy	Final Demand
W	a_1	a_2	a_3	a_4	—	—	
X	a_7	a_8	a_9	a_{10}	a_{11}	—	
Y	a_{13}	a_{14}	a_{15}	a_{16}	a_{17}		
Z	a_{19}	a_{20}	a_{21}	a_{22}	—	—	
Coal	a_{25}	a_{26}	a_{27}	a_{28}	a_{29}	—	10
Solar Energy	—	—	—	—	—	—	—

Partial Conversion

	W	X	Y	Z	Coal	Solar Energy	Final Demand
W	a_1	a_2	a_3	a_4	—	a_6	
X	a_7	a_8	a_9	a_{10}	a_{11}	—	
Y	a_{13}	a_{14}	a_{15}	a_{16}	a_{17}	—	
Z	a_{19}	a_{20}	a_{21}	a_{22}	—	a_{24}	
Coal	a_{25}	a_{26}	a_{27}	a_{28}	a_{29}	a_{30}	5
Solar Energy	a_{31}	a_{32}	a_{33}	a_{34}	a_{35}	a_{36}	5

Total Conversion

	W	X	Y	Z	Coal	Solar Energy	Final Demand
W	a_1	a_2	a_3	a_4	—	a_6	
X	a_7	a_8	a_9	a_{10}	—	—	
Y	a_{13}	a_{14}	a_{15}	a_{16}	—	—	
Z	a_{19}	a_{20}	a_{21}	a_{22}	—	a_{24}	
Coal	—	—	—	—	—	—	—
Solar Energy	a_{31}	a_{32}	a_{33}	a_{34}	a_{35}	a_{36}	10

As can be seen in the "no conversion" case, the coal industry requires inputs from industries X and Y. In the "total conversion" case, the solar energy industry will require inputs from industries W and Z. The "partial conversion" case is simply an example of an economy using both types of energy.

An important input into the above analysis is the rate at which final demand will change.

A very rough estimate could be inserted here, possibly by asking the relevant people; or an alternative course would be to simulate the entire systems for different rates of adjustment. This path would provide useful information about the effects on different industries from a shift to the SSPS as the source of power.

It must be noted here that the use of solar energy will be affected not only by a shift in final demand, but also by changes in the production technology reflected in the matrix of interindustry linkages. Changes in their matrix will also have to be considered. As individual industries shift their sources of power and as resource inputs to the energy sectors differ, the industry impacts will be reduced. These will produce new demand patterns for material, and capital and labor resources.

Along these same lines, therefore, an analysis should be presented on the demand for labor as an input to production. Since different energy-producing industries have varying labor requirements, the substitution of one for another will affect the total demand for labor in the economy. While excess labor will eventually find new jobs, there may well be specific programs which the government could institute to ease the adjustment.

CONCLUDING REMARKS AND RECOMMENDATIONS

Structure and Control Techniques

Conclusions. — State-of-the-art analytical techniques and tools are adequate for the structural and dynamic analyses needed for the SSPS structure.

Desirable material characteristics can be identified and technology developments specified to provide inputs leading to the design of structure and attitude control systems for the very large area, light-weight space structures represented by the SSPS.

The flight control performance of the SSPS baseline design can be established and parametric studies performed to determine the influence of structural flexibility upon attitude control performance. An interrelated structure/attitude control dynamic model, including such parameters as structural stiffness (frequency), steady-state attitude error, control gains and thrust levels, response time, damping ratios, and control frequencies, can be used to predict SSPS flight control performance.

The flight control performance evaluation indicated that the pointing accuracy of the SSPS fell well within the ± 1-deg limit specified by the baseline requirements for the pitch, roll, and yaw axes. In addition, the system's response time and percent overshoot were found to be acceptable for control about all three axes.

Using a digital simulation, we generated time-history response plots for the rigid-body and flexible-body dynamics of the SSPS. A comparison of the results indicated structural flexibility tended to decrease the damping characteristics of the rigid-body dynamic characteristics.

The parametric studies showed that the spacecraft's attitude errors and response times decreased as the structural frequency (stiffness) increased. For as much as a 50% decrease in structural weight (25% decrease in structural frequency) the system's pointing accuracy was still well within a ± 1-deg attitude control specification about the pitch, roll, and yaw axes. The analyses indicated that structurally the baseline design is sufficiently stiff to allow excellent attitude control.

Control of the system can be achieved to maintain the required attitude of the SSPS in orbit with a limited expenditure of propellants.

Recommendations. — A summary of efforts recommended for further study in this area follows:

- Evaluate the structure/control implications imposed by orbital assembly, including assessments of alternative assembly modes;

- Analyze the implications of thermal transients induced by eclipse periods;

- Refine structure/control interaction analyses to better understand the dynamic behavior of very large structures; and.

- Identify the minimum weight vehicle system having acceptable structural stiffness and pointing capabilities.

Before proceeding to a more refined analysis of the SSPS configuration's dynamic behavior, better definition of the vehicle structure is required. In order to do this, analyses of the implications of assembling the SSPS in orbit are needed. Specific considerations are: (a) attitude control during assembly, (b) docking closure rates and the resulting induced structural loads, (c) joining techniques for electrically conductive and non-conductive structures, and (d) structural element packaging and deployment.

In addition, refined analyses of the dynamic behavior of the spacecraft should consider: (a) the effects of rapid thermal transients, (b) local excitation of the structure by the attitude control system thrusters, (c) structural non-linearities, (d) centrifugal and centripetal effects, and (e) torsional modes and modal cross-couplings.

These efforts should ultimately be focussed on the identification of the minimum-weight spacecraft system and structure having acceptable structural stiffness and pointing capabilities.

RFI Avoidance Techniques

Conclusions. – It is possible to select a near optimum range for the fundamental frequency for the SSPS to minimize interference with other users. At the selected frequency, the design of microwave generating devices, such as the amplitron and associated filtering, can be pursued with confidence.

The RFI avoidance techniques were investigated on the basis of a model and a set of assumptions for the microwave transmission systems. The model included orbital and ground locations, ground power transmission, device characteristics, phase front control, efficiencies, induced RF environment, ionospheric and atmospheric attenuation, major frequency segments, specific frequency, typical users, and selected equipment. The noise associated with the generator, a transmitting antenna, and a filtering system placed between the antenna and the generator was found to be acceptable with other users such as troposervice, radio astronomy, and fixed microwave installations for up to 100 SSPS's. Slight RFI can be expected for shipboard radar, while amateur sharing, state police radar, and radio location from high-power defense radar will suffer substantial interference.

Recommendations. – The following tasks will require further study:

- Detail design of devices, including filters and estimates of associated costs to assure that the design goal for noise and harmonic filtering can be met within the constraints of weight and cost.

178

- Investigation of transients associated with start-up, shut-down, and physical interference or dispersion due to clouds, atmospheric and ionospheric elements, aircraft and the onset of shadowing by the Earth.

- Investigations of effects of RFI due to data, control, and command links and frequency allocation for these links.

- Acquisition of detailed data from Earth resource and ecological studies to guide receiving antenna site selection, supported by detailed data on characteristics of expected sidelobes and harmonics.

- Investigation of RFI with ship-borne radar and other spectrum users based on detailed consideration of advanced technology achievable in amplitron device design and filtering.

- Identification of alternative approaches to compensate displaced users, possibly including amateur sharing, state police radar, and radio location from high-power defense radar in the 3.23 to 3.37 GHz band.

Identification of Key Issues

Technological Issues. —

a. Microwave Generation, Transmission, and Rectification

1. Conclusions. — The microwave portion of the electromagnetic spectrum has been selected as the most useful for SSPS power generation, transmission, and rectification. For the purpose of international discussions and negotiations which will be required to obtain a frequency allocation for the SSPS, devices capable of achieving the system performance of the SSPS, but operating in other spectral regions, will have to be investigated and documented. Although requirements for high efficiency, long life, and low cost constitute significant technical challenges for development of an appropriate microwave power system, many design resources — systems, materials, and devices — already exist and are adaptable to the SSPS. Significant effort is required to achieve near optimum characteristics for the microwave generating devices, including low noise and harmonic output with high efficiency, long life, and low cost.

Control of the large-aperture microwave power beam from pointing and focusing points of view requires the phase front to be controlled with significant precision and response, such that the total antenna should be made up of many controllable subarrays, but several schemes must be investigated to determine the approach to be taken.

Illumination pattern studies for the antenna indicate moderate criticality with respect to suppression of side lobes and the recommended pattern must come from a progressive series of in-depth studies involving the structure, Amplitron, subarray, phase front control and biological effects of side lobes.

Successful demonstration of high-efficiency rectification of microwaves to dc is expected to show that the SSPS will be capable of generating power on Earth with an efficiency which has not yet been equalled by any known power generation method.

Although the materials and electrical problems associated with the development of the microwave system, particularly in the case of the spaceborne power-transmitting antenna and its control system are considered formidable, we feel they are solvable on the basis of known physical sciences. Thus, we feel that engineering, research, and development, rather than any scientific breakthrough, represent the keys to their solution.

2. Recommendations. – The technologies for microwave power generation, transmission, rectification and control from synchronous orbit to the Earth will require:

- Demonstrations of high efficiency for dc-to-dc power transmission;

- Investigation of approaches to implementing the adaptive array principle;

- Design construction and evaluation of an Amplitron with associated filtering, and

- Definitive illumination pattern investigations.

We recommend that the Amplitron (~5 KW) be investigated further as the device capable of converting dc to microwave energy, because of the simplicity and long life inherent in the pure metal cold cathode and its compatibility with simple passive waste heat rejection schemes.

b. Solar-Energy Conversion

1. Conclusions. – Design approaches for the solar collector, solar cell blankets, and power collection and distribution methods have been evolved to meet the requirements of the structure and control technique analyses. The possible variation in power output caused by such effects as solar collector and blanket distortions, surface degradation, and attitude control was found to be within design limits. Possible power interruptions caused by meteoroid impacts, shadowing, internal damage, and/or temperature runaway conditions were considered. Based on these considerations, key issues, and performance goals can be established, and for each of the key issues, distinct areas can be identified, development objectives listed, and the approach indicated. The advanced state-of-the-art of photovoltaic solar energy conversion is indicated by the detailed development programs which can be identified for this portion of the SSPS system.

2. Recommendations. – An integrated technology development and verification program should be undertaken for the SSPS solar energy conversion components and subsystems in conjunction with ongoing or planned NASA and NSF programs.

Environmental Issues. –

a. Environmental/Ecological Impact

1. Conclusions. – The environmental and ecological impacts of the SSPS include the following:

⊖ Waste heat released at the receiving antenna does not constitute a significant thermal effect on the atmosphere. If the antenna is located in desert regions where water is limited, there may be some slight modification of the plant community at and near the antenna site.

⊚ With RF shielding incorporated below the receiving antenna, operation can be compatible with other land uses, because there is only a small degree of reduction of solar radiation received on the ground below the antenna. However, installation and maintenance of the antenna has to be planned because extensive activities may be damaging to some ecologically important systems.

⊛ Injection of water into the stratosphere and upper atmosphere by space vehicle exhausts will be small in contrast to the natural abundance.

2. Recommendations. – More information on upper atmosphere properties and processes occuring there should be compiled so the effects of space shuttle flight water and NO injection can be established in detail.

b. Biological Effects

1. Conclusions. – The SSPS can be designed to accommodate a wide range of microwave power flux densities to meet internationally accepted standards of microwave exposure. The transmitting antenna size, the shape of the microwave power distribution across the antenna, and the total power transmitted will determine the level of microwave power flux densities in the beam reaching the Earth.

Precise pointing of the microwave beam can be achieved with attitude stabilization and automatic phase control to assure efficient transmission of the power to the receiving antenna. The design approaches already identified indicate that this objective can be met.

As to the biological effects of long-term exposure to microwaves, research plans developed and recommended by the Office of Telecommunications Policy (OTP) are addressing the relatively known adverse biological effects due to the generation of heat by microwaves and other radio frequency radiation at high intensities, as well as the extent and importance of inadequately known, but more subtle, changes which may occur at lower intensities.

Recognizing that an assessment of biological effects of non-ionizing electromagnetic radiation for the whole spectrum will be provided through the OTP, we feel that the SSPS will not present further significant or unique biological effects on the public due to the microwave radiation.

2. Recommendations. – The effects on birds exposed to microwave power flux densities within the beam at the receiving antenna and the effects on aircraft flying through the beam, even though projected to be negligible, should be determined experimentally.

An overall SSPS-related biological effects program should be carried out and closely integrated with the research plans for a national biological effects program being developed by the Office of Telecommunications Policy.

Economic Issues. —

1. Conclusions. — The cost projections available for the SSPS components and systems indicate that there is a reasonable probability that the SSPS can become competitive with other power-generating methods as technology developments proceed.

There are also clear indications that SSPS costs can be reduced if a program of mass manufacture is undertaken in support of SSPS development.

An economic comparison of the SSPS with other means of generating power and the methodology which can be made to deal with these issues would include:

- The costs and benefits associated with the SSPS which have to be evaluated to determine the economic feasibility of an investment of this type;

- The macro-economic interindustry effects produced by the SSPS which have to be examined to analyze the effects such an investment might have on the structure of the economy as a whole.

- The consumption effects created by the SSPS which will be reflected in both the cost/benefit analysis and the analysis of macro-economic interindustry effects.

These key issues can be related not only to an efficiency criterion where an optimal allocation of scarce resources is the sole objective, but also to indicate the effects associated with large-scale changes that might be created in the economic structure if the projected benefits of an SSPS are to be realized.

2. Recommendations. — A technology assessment, including technical, environmental, economic, and social factors, should be carried out to provide data for a decision-making process among different energy-production systems required to meet future energy demands.

Program Phasing

1. Conclusions. — Based on an assessment of the steps required to develop the various technologies for the SSPS, three major program phases can be identified, as shown in Figure 92. The technology development and verification program should achieve the following primary goals:

- Be ready — if necessary — to initiate SSPS prototype system development; and

- Provide tangible returns to the public in the event the option to proceed is *not* exercised.

182

Fullfillment of these goals will require a better definition of the SSPS prototype's desirable performance, size, and estimated cost. Once these have been broadly established, a finer grained structure of technology and cost goals can be folded into present technology development and verification plans.

Present studies of an operational SSPS have identified those technical areas where impressive performance and cost gains can be expected. These are:

- Efficient solar energy conversions at elevated temperatures;

- Long range microwave power transmission; and

- Assembly and control of large semiflexible spacecraft structures.

Within each of these broad categories, specific performance levels and weight and cost goals have been generally identified for a baseline, operational SSPS design. These represent the objective framework that, when integrated with prototype SSPS needs, establish detailed goals for the technology development and verification activity.

2. <u>Recommendations</u>. – To achieve these goals, a variety of analyses, ground tests, and development flight activities should be carried out. Therefore, as a next step, a Phase A-type study should be initiated to establish the most cost-effective strategy for achieving these goals. The resulting program would culminate with shuttle-borne payloads that would verify developed concepts; conversely, early definition of these payloads would provide the near-term focus for the technology development and verification program.

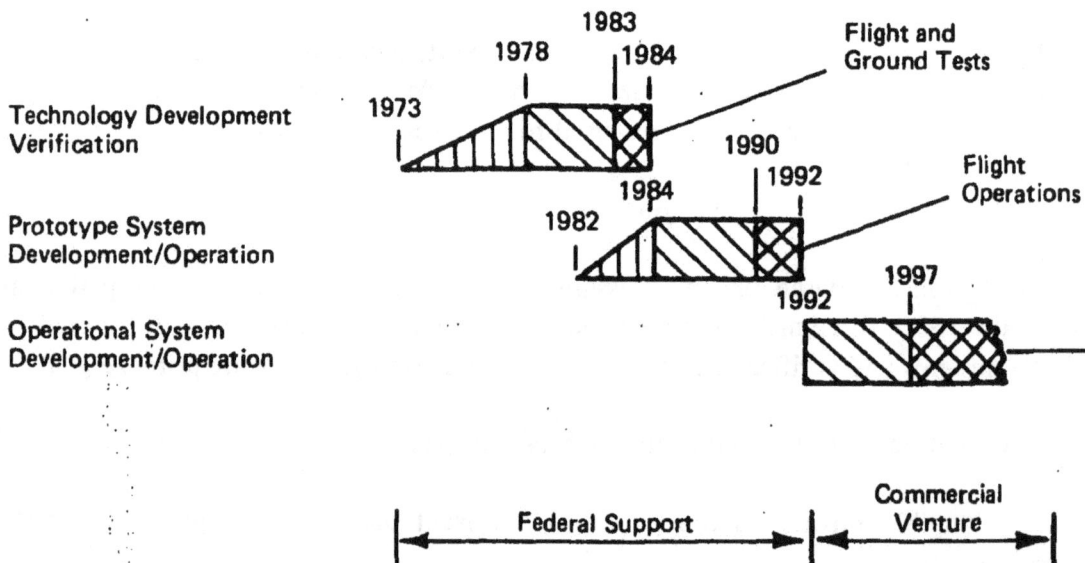

FIGURE 92. – PROGRAM PHASING

183

REFERENCES

1. Glaser, P.E., Solar Energy as a National Energy Resource, NSF/NASA Solar Energy Panel, Dept. of Mechanical Engineering, University of Maryland, College Park, Md., December 1972.

2. Glaser, P.E., The Future of Power from the Sun, IECEC 1968 Record; IEEE Publication 68C21-Energy, 1968, pp. 98-103.

3. Glaser, P.E., Power from the Sun: Its Future, Amer. Assn. Advan. Sci., Vol. 162, 22 November 1968, pp. 857-61.

4. Solar Satellite Power Station, Tech Memoranda, Grumman Aerospace Corporation, Bethpage N.Y., January-June 1972.

5. Satellite Solar Power Station, Technical Report, Q-71098, Spectrolab/Heliotek Divisions, Textron, Inc., November 1971.

6. Satellite Solar Power Station, Arthur D. Little, Inc., January 21, 1972.

7. Satellite Solar Power Station, Configuration Status Report, Grumman Aerospace Corporation, ASP-583-R-10, June 1972.

8. Satellite Solar Power Station, Master Program Plan Development, Grumman Aerospace Corporation, ASP-611-R-12, June 1972.

9. Microwave Power Transmission in the Satellite Solar Power Station System. Technical Report ER72-4038, Raytheon Company, 27 January 1972.

10. "Briefings before the Task Force on Energy of the Subcommittee on Science, Research and Development of the Committee on Science and Astronautics," U.S. House of Representatives, 92nd Congress, Second Session, Series Q, March 1972, U.S. Government Printing Office, Washington, 1972.

11. Wolf, M., Cost Goals for Silicon Solar Arrays for Large-Scale Terrestrial Applications, Conference Record of the Ninth IEEE Photovoltaic Specialists Conference, Silver Spring, Maryland, May 1972, pp. 342-350.

12. Berman, P.A., Photovoltaic Solar Array Technology Required for Three Wide-Scale Generating Systems for Terrestrial Applications: Rooftop, Solar Farm, and Satellite, Technical Report 32-1573, California Institute of Technology, Jet Propulsion Laboratory, Pasadena, California, October 15, 1972.

13. Lindmayer, I., and Allison, J., An Improved Silicon Solar Cell — The Violet Cell, Conference Record of the Ninth IEEE Photovoltaic Specialists Conference, Silver Spring, Maryland, May 1972.

14. "Solar Cells, Outlook for Improved Efficiency," Ad Hoc Panel on Solar Cell Efficiency, National Research Council, National Academy of Sciences, Washington, D.C., 1972.

15. Chadda, T.B.S., and M. Wolf, "The Effect of Surface Recombination Velocity on the Performance of Vertical Multi-Junction Solar Cell," Conference Record of the Ninth IEEE Photovoltaic Specialists Conference, Silver Spring, Maryland, May 1972, pp. 87-90.

16. Woodall, J.M., "Conversion of Electromagnetic Radiation to Electrical Power," Pat. 3,675,026, issued July 4, 1972.

17. Gutman, F., and L.E. Lyons, *Organic Semiconductors*, Wiley, New York, 1967.

18. Brown, W.C., Experiments in the Transportation of Energy by Microwave Beams, 1964 IEEE Intersociety Conference Record, Vol. 12, Pt. 2, 1964, pp. 8-17.

19. Goubau, G., Microwave Power Transmission from an Orbiting Solar Power Station, *Microwave Power*, Vol. 5, No. 4, December 1970, pp. 223-231.

20. Falcone, V.J., Jr., Atmospheric Attenuation of Microwave Power, *Microwave Power*, Vol. 5, No. 4, December 1970, pp. 269-278.

21. Brown, W.C., The Satellite Solar Power Station, IEEE Spectrum, March 1973.

22. Brown, W.C., High-Power Microwave Generators of the Crossed-Field Type, *Microwave Power*, Vol. 5, No. 4, December 1970, pp. 245-259.

23. Glaser, P.E., The Potential of Power From Space, 1972 IEEE EASCON Record, pp. 34-41.

24. Brown, W.C., op. cit., March 1973, pp. 38-47.

25. Special Issue on Active and Adaptive Antennae, IEEE Transactions of Professional Group of Antennae and Propagation, March 1964.

26. Brown, W.C., The Receiving Antenna and Microwave Power Rectification, *Microwave Power*, Vol. 5, No. 4, December 1970, pp. 279-292.

27. IDEAS Manual, Volume IIB, Grumman Aerospace Corporation.

28. NASTRAN User's Manual (NASA SP-222).

29. Selection of the Baseline Attitude Control System for the SSPS and a Stability and Performance Analysis of the Elastic Coupling between the Control System and the Spacecraft's Structural Modes, ASP-611-M-1009, Grumman Corporation, 21 September 1972.

30. Performance Evaluation and Parametric Sizing Study of the Baseline SSPS, ASP-611-M-1019 Grumman Corporation, 2 January 1973.

31. Sensitivity of Attitude Control Propellant Requirements to SSPS Deviation Angle Limits, ASP-611-M-1004, Grumman Corporation, 21 August 1972.

32. SSPS Electrical Power Transmission and Distribution System — Sizing Data and Internal Electro-Magnetic Forces, ASP-611-M-1007, Grumman Corporation, 23 August 1972.

33. Feasibility Study of Satellite Solar Power Station (SSPS), Spectrolab Report 6011-03, 4 October 1972.

34. Limitation on SSPS Control Forces So That Structural Rotations Will Not Exceed Satellite Pointing Accuracy of ± 1°, ASP-611-M-1018, Grumman Corporation, 6 December 1972.

35. Selection of the Baseline Attitude Control System for the SSPS and a Stability and Performance Analysis of the Elastic Coupling between the Control System and the Spacecraft's Structural Modes, ASP-611-M-1009, Grumman Corporation, 21 September 1972.

36. Structural and Dynamic Analysis of the Satellite Solar Power Station, ASP-611-M-1020, Grumman Corporation, 2 January 1973.

37. SSPS Microwave Transmission System Model and Assumptions, Engineering Memorandum, File BII-P, Raytheon Company, Wayland, Mass., August 28, 1972.

38. Report ER-72-4038, Section 4.3.3.2.5, Raytheon Company, Wayland, Mass.

39. Falcone, V.J., Jr., Atmospheric Attenuation of Microwave Power, J. Microwave Power, Vol. 5, No. 4, December 1970.

40. Valley, S.L., ed., Handbook of Geophysics and Space Environments, AFCRL Office of Aerospace Research, USAF 1965.

41. Medhurst, R.G., Comparison of Theory and Measurement, IEEE Transactions on Antenna Preparations, July 1965.

42. Memorandum for Record LKI/2954 RCS No. 2-18, Satellite Solar Power Station Study, 20 October 1972. Visit Report — Owen Maynard of Raytheon with N. Sissenwine, A. Kantor, I. Salmeda, and D. Granthan of Design Climatology Branch, Aeronomy Laboratory. Air Force Cambridge Research Laboratories (AFCRL), L.G. Hanscom Field, Bedford, Mass.

43. Lenhard, R.W., Cole, A.E., and Sissenwine, N., Environmental Paper No. 350, Preliminary Models for Determining Instantaneous Precipitation Intensities from Available Climatology, AFCRL-71-0168, 5 March 1971.

44. Local Climatological Data, U.S. Department of Commerce, National Oceanic & Atmospheric Administration Environmental Data Service, Washington, D.C., 1972. Also National Oceanographic & Atmospheric Agency, National Weather Service H.Q. in Washington, National Weather Center in Asheville, N.C., and Hasschbarger, H.R. (National Oceanographic Environmental Data Service), Head of Climatology, located in Gramax Building, Silver Spring, Maryland.

45. Salmeda, H.A., Sissenwine, N., and Lenhard, R.W., Environmental Research Paper No. 374, Preliminary Atlas of 1.0, 0.5, and 0.1 Percent Precipitation Intensities for Eurasia. AFCRL-71-0527, 7 October 1971.

46. Barnes, A.A., Jr., Atmospheric Water Vapor Divergence: Measurements and Applications, AFCRL-65-501, Special Reports No. 28, July 1965.

47. Granthan, D.D., and Kantor, A.J., Distribution of Radar Echoes over the United States, Air Force Surveys in Geophysics No. 171, AFCRL-67-0232, April 1967.

48. Raytheon Equipment Division, Special Projects Office, Engineering Memorandum OEM: 72:15, File No. B II-P, SSPS Microwave Transmission System Model and Assumptions, August 28, 1972.

49. Budyko, M.T., 1955 *Atlas of the Heat Balance,* Leningrad, 1955, and *The Heat Balance of the Earth's Surface.* U.S. Department of Commerce, Office of Technical Services, Washington, D.C., 1965, also 1971 *Climate and Life,* Hydrological Publishing House, Leningrad, 1971.

50. Gates, D.M., *Energy Exchange in the Biosphere.* Harper and Row, New York 1962.

51. Major, J., Potential Evapotranspiration and Plant Distribution in Western States with Emphasis on California, pp. 93-126 in Shaw, R.H. (ed.) "Ground Level Climatology" AAAS, Washington, D.C. 1967, p. 395.

52. Landsberg, H.E., Climates and Urban Planning in *Urban Climates,* WMU, Geneva, 1970, p. 372.

53. Huff, F.A., and Chagnon, S.A., Jr., Climatological Assessment of Urban Effects on Precipitation at St. Louis, J. Appl. Met., Vol. II, 1972, pp. 823-842.

54. SMIC, Inadvertent Climate Modification, Report of the Study of Man's Impact on Climate, MIT Press, Cambridge, Mass., 1971.

55. Martell, A.E., Residence Times and Other Factors Influencing Pollution of the Upper Atmosphere, in *Man's Impact on the Climate*, Mathews, W.H., Kellogg, W.W., and Robinson, G.D. (eds.), MIT Press, Cambridge, Mass., 1971, p. 594.

56. Johnston, H., (1971) *Sci.*, Vol. 173, 1971, p. 517, and The Role of Chemistry and Air Motions on Stratospheric Ozone as Affected by Natural and Artificial Oxides of Nitrogen, Paper presented at Autumn Meeting of NAS, Washington, D.C., 1971.

57. Johnston, H., Laboratory Chemical Kinetics as an Atmospheric Science, 1972, pp. 90-114, in Climatic Impact Assessment Program: Proceedings of the Survey Conference, Barrington, A.E. (ed.), February 15-16 1972, U.S. Department of Transportation, DOT-TSC-OST-72-13.

58. SCEP, Report on the Study of Critical Environmental Problems, MIT Press, Cambridge, Mass., 1970.

59. Crutzen, P.J., On Some Photochemical and Meteorological Factors Determining the Distribution of Ozone in the Stratosphere: Effects of Contamination by NO_x emitted from Aircraft, Report AP-6, Institute of Meteorology, University of Stockholm, 1971.

60. Nicolet, M., and Vergison, E., Aeronomica Acta, Vol. 90, 1971.

61. McElroy, M., and McConnell, J.C., J. Atmos. *Sci.* Vol. 28, 1971, p. 1095.

62. Currin, C.G., et al., Feasibility of Low-Cost Silicon Solar Cells, Conference Record of the Ninth IEEE Photovoltaic Specialists Conference, Silver Spring, Maryland, May 1972, pp. 367-369.

63. Satellite Solar Power Station, Technical Report Q-71098, Spectrolab/Heliotek Division, Textron, Inc., November 1971.

64. Bates, H.E., et al., The Edge-Defined Film-Fed Growth of Silicon Crystal Ribbon for Solar Cell Applications, Conference Record of the Ninth Photovoltaic Specialists Conference, Silver Spring, Maryland, May 1972, p. 386.

65. Currin C.G., et al., op. cit., 1972, p. 363.

66. *Perspectives on Experience*, The Boston Consulting Group, Inc., Boston, Massachusetts, 1968.

67. Berman, P.A., op. cit., October 15, 1972.

68. Salkeld, R., Mixed Mode Propulsion for the Space Shuttle, Astronautics and Aeronautics, August 1971.

69. Salkeld, R., and Beichel, R., Reusable One-Stage-To-Orbit Shuttles: Brightening Prospects, Astronautics and Aeronautics, June 1973.

70. Brown, W.C., Status of the Cost and Technology of the Wave Power Transmission System in the SSPS, Raytheon Company, PT-3738, March 13, 1971.

71. The Boston Consulting Group, Inc., Boston, Mass., 1968.

72. Buchin, S.I., Computer Programs for the Analysis of Complex Decision Problems, Harvard University, C-40RC, January 1969.

*"The aeronautical and space activities of the United States shall be
conducted so as to contribute . . . to the expansion of human knowledge of phenomena in the atmosphere and space. The Administration
shall provide for the widest practicable and appropriate dissemination
of information concerning its activities and the results thereof."*
—NATIONAL AERONAUTICS AND SPACE ACT OF 1958

NASA SCIENTIFIC AND TECHNICAL PUBLICATIONS

TECHNICAL REPORTS: Scientific and
technical information considered important,
complete, and a lasting contribution to existing
knowledge.

TECHNICAL NOTES: Information less broad
in scope but nevertheless of importance as a
contribution to existing knowledge.

TECHNICAL MEMORANDUMS:
Information receiving limited distribution
because of preliminary data, security classifica-
tion, or other reasons. Also includes conference
proceedings with either limited or unlimited
distribution.

CONTRACTOR REPORTS: Scientific and
technical information generated under a NASA
contract or grant and considered an important
contribution to existing knowledge.

TECHNICAL TRANSLATIONS: Information
published in a foreign language considered
to merit NASA distribution in English.

SPECIAL PUBLICATIONS: Information
derived from or of value to NASA activities.
Publications include final reports of major
projects, monographs, data compilations,
handbooks, sourcebooks, and special
bibliographies.

TECHNOLOGY UTILIZATION
PUBLICATIONS: Information on technology
used by NASA that may be of particular
interest in commercial and other non-aerospace
applications. Publications include Tech Briefs,
Technology Utilization Reports and
Technology Surveys.

Details on the availability of these publications may be obtained from:

SCIENTIFIC AND TECHNICAL INFORMATION OFFICE

NATIONAL AERONAUTICS AND SPACE ADMINISTRATION
Washington, D.C. 20546

NATIONAL AERONAUTICS AND SPACE ADMINISTRATION
WASHINGTON, D.C. 20546

OFFICIAL BUSINESS
PENALTY FOR PRIVATE USE $300

SPECIAL FOURTH-CLASS RATE
BOOK

POSTMASTER :   If Undeliverable (Section 158
                Postal Manual) Do Not Return